Distributed Intelligence

Problem-Solving in Slime Molds

Team Trex

Published by NobleTrex Press

For permissions and other inquiries, write to:
P.O. Box 3132, Framingham, MA 01701,
USA

Contents

Introduction

Imagine, if you will, a civil engineer tasked with designing the rail network for the Greater Tokyo Area. This is a staggering logistical challenge. The engineer must connect thirty-six geographically disparate towns to the central hub, navigating mountains and lakes, all while balancing two competing and contradictory demands: the network must be efficient (short tracks, low cost) and it must be resilient (if one line breaks, the whole system shouldn't collapse). To solve this, human engineers spend decades refining blueprints, utilizing supercomputers, and relying on generations of urban planning theory.

Now, imagine that same engineer is a bright yellow, throbbing puddle of slime. It has no eyes to see the map, no ears to hear the commuter complaints, and, most crucially, no brain to calculate the route. It is a single cell.

In 2010, a team of researchers led by Toshiyuki Nakagaki decided to pit this organism-*Physarum polycephalum*-against the complexities of Japanese urban infrastructure. They placed oat flakes (the slime mold's equivalent of a Michelin-star meal) on a wet surface at locations corresponding to the major cities surrounding Tokyo. They planted the slime mold in the center. Over the course of a day, the organism grew, fanned out, and retracted, transforming from a shapeless blob into a complex lattice of veins. When the researchers compared the slime mold's final structure to the actual Tokyo rail system, the resemblance was uncanny. The slime had recreated, with mathematical precision, a network that human engineers had spent a century optimizing. It had found the sweet spot between cost and resilience, all without a single neuron firing.

This book is about that yellow puddle. But more importantly, it is about a radical shift in how we understand intelligence itself.

For too long, we have suffered from what biological philosophers might call "cerebral chauvinism." We look for intelligence in the furrowed brows of grandmasters, the silicon chips of supercomputers, and the convoluted gray matter inside our own skulls. We assume that to solve a problem, one needs a central processor-a boss, a captain, a brain-to take in data, crunch the numbers, and issue a command. We view intelligence as a top-down hierarchy.

Physarum polycephalum shatters that hierarchy. It offers us a glimpse into a world of distributed intelligence, where "thinking" is not a specific activity performed by a specific organ, but a physical property of matter itself. This organism demonstrates that you do not need a brain to have a mind. You do not need a central nervous system to make rational economic decisions, navigate labyrinths, remember where you have been, or anticipate the future. All you need is a network of fluid, a rhythm, and a set of simple, local rules that give rise to staggering global complexity.

Why should we care about the cognitive abilities of a forest-dwelling mold? Because the problems *Physarum* solves are the same problems we face. The internet is a network that must balance efficiency and resilience, just like the slime mold's veins. Our power grids, our supply chains, and our traffic systems are all variations of the same mathematical puzzles that this organism has been solving for millions of years. As we struggle to engineer self-driving cars and swarm robots that can operate without a central controller, we are finding that nature has already written the code. We just need to learn how to read it.

To understand how a bag of chemicals can act like a genius, we must first descend into the damp, dark undergrowth where *Physarum* thrives. This organism is a biological outlaw. It defies the neat categorization of high school biology classes. It is not a plant, though it can grow; it is not a fungus, though it loves damp logs; and it is not an animal, though it hunts. It is a protist, a true oddity of the natural world. In its "plasmodium" stage-the active, thinking phase-it is a single cell that can grow to the size of a bathmat, containing millions of nuclei floating freely in a shared cytoplasm.

If you were to look at *Physarum* under a microscope, you wouldn't see static tissue. You would see a river. The organism is in a state of perpetual, rhythmic flow. This is the heartbeat of the machine, a mechanism known as shuttle streaming. The cytoplasm surges back and forth, pushed by contracting tubes, carrying nutrients and information from one end of the cell to the other. It is a hydraulic computer, calculating through flow.

This rhythmic pulsation is the engine of its intelligence, and as we analyze the physics of this "liquid lattice," we begin to see how the magic happens. The organism is an expert at resource allocation. By expanding veins that carry a heavy load of nutrients and shrinking those that don't, the slime mold physically embodies the principle of "use it or lose it." It is a living graph, constantly updating its own geometry to match the reality of its environment. This dynamic adaptation allows it to solve the "Shortest Path Problem," a classic challenge in computer science. Put the slime in a maze with food at the start and end, and it will retract its dead-end branches until only the single, most efficient route remains. It does not calculate the path; it *becomes* the path.

But the story gets stranger. As we probe deeper into the capabilities of this brainless thinker, we find behaviors that feel uncomfortably human. We will examine experiments where *Physarum* makes difficult trade-offs. When presented with high-quality food in a dangerous, illuminated environment versus low-quality food in the safety of the dark, the slime mold weighs the risk against the reward. It calculates the "marginal utility" of the meal. It hesitates. It deliberates. And it makes a choice that mirrors the rational decision-making models used by economists.

We will also uncover the slime mold's memory. Lacking a hippocampus to store dates and facts, *Physarum* offloads its memory into the environment. As it crawls, it leaves behind a translucent trail of non-living slime. When it encounters this trail again, it senses that it has already explored that area and turns away. This is a spatial memory written in carbohydrates, a "Hansel and Gretel" strategy that prevents the organism from swimming in circles. It forces us to ask: Is memory something stored inside us, or is it a relationship between a body and its world?

The implications of this research extend far beyond biology. Computer scientists are now deriving "Physarum Solvers"-algorithms based on the mold's growth rules-to tackle complex computational hurdles. Engineers are looking to the slime mold to design "soft robots," amorphous machines that can squeeze through rubble to find earthquake survivors, guided not by a silicon chip, but by the fluid dynamics of their own bodies. We are witnessing the birth of a new kind of engineering, one that moves away from rigid, centralized control toward flexible, self-organizing systems.

Yet, there is a philosophical dimension here that is equally compelling. When we watch a time-lapse video of *Physarum* pulsating, expanding, and solving problems, we are confronted with the "alien" nature of intelligence. We are so used to thinking of consciousness as a spotlight located behind our eyes. But intelligence might be more like a tide-a pervasive, distributed force that emerges wherever matter organizes itself to survive.

Throughout these pages, we will walk alongside the pioneers of this field. We will see how early curiosity in the late 19th century transformed into the rigorous biophysics of today. We will step inside the labs where scientists use time-lapse macro-photography to speed up the slow-motion drama of the slime mold's life, revealing a dynamism invisible to the naked eye. We will look at the data-the graph theory and the flow dynamics-that prove this isn't just random growth. It is optimization.

We will also confront the limits of the mold. It is not infallible. It can be tricked. It can be overwhelmed. By understanding where *Physarum* fails, we learn about the boundaries of heuristic problem-solving and the specific advantages of having a centralized brain. The goal is not to prove that slime molds are "smarter" than us, but to understand that they are smart in a fundamentally different way-a way that we are only just beginning to comprehend.

This investigation challenges our ego. It suggests that the ability to process information, to value one outcome over another, and to navigate a complex world is not a special privilege of creatures with spines and skulls. It is a fundamental capacity of life. As we unravel the mechanisms of the slime mold, we are effectively taking apart the engine of cognition to see how the gears turn when there is no driver at the wheel.

So, forget what you think you know about thinking. Forget the image of the computer processor or the firing neuron. Let us look instead at the pulsing yellow vein, the flow of the fluid, and the silent, rhythmic wisdom of the slime. The smartest thing you encounter today might just be growing on a rotting log in your backyard. Let's find out how it works.

Chapter 1

The Brainless Thinker

Imagine a creature that maps transit networks and navigates labyrinths, yet possesses neither neurons nor a brain. This chapter introduces Physarum polycephalum, a single-celled biological anomaly that forces us to dismantle and reconstruct our understanding of what it means to be intelligent.

1.1 The Million-Nuclei Cell

A slime mold in the genus *Physarum* can look like a small accident of nature: a smear of yolk-yellow paint, a spilled drop of custard, a scrap of damp sponge left too long on wood. Then it starts to move. Not quickly, not with the purposeful gait of an animal, but with an unmistakable insistence, as if the thing is *deciding* where to go and then slowly committing its whole body to the choice. The surprise is not only that it moves, but that it does so without the architectural features we use, almost by reflex, to explain coordinated behaviour. No brain. No ganglia. No nerves. Not even the usual cellular boundaries that make most organisms legible to our intuitions.

The form that produces this unease—and this fascination—is the plasmodium. It is a feeding stage: broad, creeping, opportunistic, happiest in darkness and moisture, drawn to the bacterial and fungal films that coat rotting leaves and decaying logs. Botanists and naturalists encountered it long before it became a laboratory celebrity, and their descriptions have the mixture of precision and disbelief that often marks first contact with an organism that refuses to sit comfortably inside a category. If you are expecting a plant, it is too mobile. If you are expecting an animal, it is too amorphous. If you are expecting a fungus, it has the wrong kind of body. The plasmodium behaves less like a tidy individual and more like a living process that happens to have a membrane.

The central fact, the one that makes almost everything else about this organism possible, is also the one that most offends our sense of what a cell is. The plasmodium is a single cell. Not "single" in the everyday sense of small and self-contained, but single in the literal sense: one continuous volume of cytoplasm enclosed by one plasma membrane. Inside that membrane are not one nucleus or a neat handful, but a vast number of nuclei sharing the same cytoplasm. They are carried around like passengers in a common sea, dividing and drifting in a body that can grow to fill a Petri dish and, in the wild, can spread much larger.

This is a syncytium: a multinucleate cell. The term sounds abstract until you picture what it entails. In an animal, the nucleus is a kind of headquarters. It holds the DNA, coordinates gene expression, and in a typical cell it also anchors the story of individuality: one genome in one nucleus in one cell. *Physarum* keeps the genome in many nuclei and lets the rest of the cell become a shared infrastructure. Imagine a city in which every neighbourhood has its own town hall with copies of the laws and records, but the roads, water, and electricity form one continuous network with no hard borders. You can still ask what the city is doing, where it is going, what it is building—but the answers will not be found in a single central office.

The number of nuclei is often described in extravagant terms, and for good reason. In laboratory descriptions, it is common to encounter references to millions or even billions of nuclei in a mature plasmodium. The exact number depends on size and physiological state, but the crucial point is qualitative: the plasmodium is not a cell that happens to have a few extra nuclei. It is a cell whose basic operating principle is *many nuclei, one body.* Those nuclei do not live in splendid isolation either. They can divide in synchrony, as if the whole organism has a shared sense of time at the level of the cell cycle. That synchrony is not merely a curiosity; it is a clue that signals and materials are being distributed across the organism with unusual efficiency. A plasmodium can coordinate itself, at scale, without the apparatus of a circulatory system or a nervous system. It is a giant cell that behaves like an integrated animal.

How does such a cell hold itself together? Part of the answer is mechanical. The plasmodium is not a uniform blob; it is an evolving network of veins. The thicker veins act as highways, the thinner ones as local streets, and the advancing edges spread as fans and sheets over the surface. If you watch a well-fed plasmodium on agar, you see something like a living map being redrawn: tubes appear, thicken, shrink, vanish; edges extend, hesitate, retract; the whole organism subtly changes its geometry as if it were testing alternative plans and then committing to one.

The more important part of the answer is dynamic. Inside those veins, the cytoplasm does not sit still. It streams.

If you take the organism seriously as one cell, cytoplasmic streaming is more than an interesting motion; it is the organism's internal logistics. The streaming in *Physarum* has a distinctive pattern called shuttle streaming. "Shuttle" is the right word because the flow reverses direction rhythmically, back and forth, like a tide in a narrow channel. For a period the cytoplasm surges one way; then, after a brief pause, it runs back. Over time, there can be net transport because the organism's shape and contractions are not perfectly symmetric, but the striking feature is the oscillation itself: a body-wide pulse of movement that does not require a heart, because the entire body behaves as the pump.

The mechanism is rooted in contraction. The plasmodium's cortex—the thin layer near the membrane—contains actin and myosin, the same molecular machinery used in animal muscle. In *Physarum*, waves of contraction and relaxation travel through the network. When a vein segment contracts, pressure pushes cytoplasm through the tube; when it relaxes, the pressure gradient changes and flow reverses or redistributes. Chemical signals, including calcium dynamics, are intimately tied to this contractile rhythm. The result is an organism whose "circulation" is inseparable from its "posture." In an animal, the muscles move the body and the heart moves the blood. In *Physarum*, movement, feeding, and internal communication share the same physical engine.

This matters because a giant cell is, in principle, a problem. A typical cell is small partly because diffusion is slow. If a molecule has to travel across a centimetre by random jostling, it takes too long to be useful. The plasmodium solves the scaling problem by turning its whole interior into a stirred, pumped medium. Streaming mixes nutrients, distributes metabolites, and carries signals. It also couples distant parts of the organism: an event at one edge can, through changes in flow and contraction patterns, influence behaviour elsewhere. The organism is not merely large; it is *large and coherent.*

A nineteenth-century observer, encountering this pulsing yellow network in a damp woodland, might not have had the vocabulary of actomyosin and calcium waves, but could still register the uncanny unity of the thing. The plasmodium does not behave like a collection of independent amoebae. It behaves like one animal spread into a sheet. This is why early classification was so tortured. It looks fungal when it produces sporangia, amoeboid when it moves, plantlike when it sprawls. Even the names carry traces of uncertainty. "Slime mould" is a compromise phrase, both descriptive and misleading. "Myxomycete" (slime fungus) nods toward an older fungal affiliation. Modern taxonomy places *Physarum* among the amoebozoans rather than fungi, but the deeper point is that it forces taxonomists, and the rest of us, to distinguish between superficial similarities and underlying organisation.

In the twentieth century, *Physarum polycephalum* (the species name itself, "many-headed," seems almost too eager to make the point) became a standard organism in cell biology labs for reasons that now seem obvious. The plasmodium is large enough to handle with tweezers. It tolerates cutting and re-fusing. It can be grown on simple media and fed with humble foods like oat flakes. It is translucent enough that internal processes can be watched. Many experiments that would be delicate microsurgery in a typical cell become, in *Physarum*, something closer to carpentry: you can carve channels, place barriers, add attractants, and then watch how the organism rearranges its body.

A concrete example makes this tactile. A researcher can place a piece of plasmodium at one end of a damp agar plate and a line of oat flakes at the other. The organism does not ooze uniformly toward the food like syrup running downhill. It extends exploratory fronts in multiple directions, then stabilizes thicker veins along the routes that prove rewarding. Under time-lapse, the plate becomes a kind of theatre for distributed decision-making. But the physical foundation of that decision-making is already visible here, before any talk of "intelligence": the organism is continually pumping itself, continually redistributing its own substance. It is a self-remodelling hydraulic network.

The question that naturally follows is how the many nuclei fit into this. Do the nuclei have specialised roles, like organs? Are they different in different regions, like the cells of a multicellular organism? The answer is both simpler and more radical. The nuclei in a plasmodium are not separate cell types. They are, broadly, copies of the same genome in a shared cytoplasm. The organism's differentiation, such as it is, happens through local physiology rather than hard cellular boundaries. The network can have regions that behave differently—a vigorously advancing front, a thick transport vein, a shrinking branch—without those regions being made of different kinds of cells. They are different *states* of the same continuous cell.

This helps explain why the plasmodium can be so plastic. In an animal, changing the body's plan requires coordinating many cells, each with its own membrane and local constraints. In *Physarum*, changing the plan can be as immediate as shifting contractile rhythms and flow patterns, followed by reinforcement of some tubes and abandonment of others. The nuclei supply the genetic capacity, but the body supplies the computation in a more physical sense: pressure, viscosity, elasticity, and chemistry become the medium of control.

One of the most illuminating aspects of the plasmodium is how it makes boundaries and then dissolves them. If you take two pieces of plasmodium from the same strain and place them near each other on a plate, they can fuse. The membrane contact becomes a bridge, and then the bridge becomes a shared cytoplasm. The organism can literally become one with itself again after being separated. This ability undercuts another intuition we often carry without noticing: that an individual is something bounded and persistent. For *Physarum*, individuality is a matter of continuity of flow and coordination, not a sacrosanct shape.

A historical anecdote captures both the simplicity of the laboratory practice and the strangeness of the biology. In the era when *Physarum* was becoming a workhorse for studying cell cycles and differentiation, researchers took advantage of the organism's size to perform microinjections and manipulations that would have been nearly impossible in smaller cells. The plasmodium could be punctured and would heal; substances could be introduced into the flowing cytoplasm and would be carried, by the organism's own streaming, across substantial distances. What looks, to a casual observer, like a fragile smear is in fact robust soft matter, engineered by evolution to cope with cuts, drying edges, and the irregular terrain of leaf litter. It is difficult not to feel that the plasmodium is a living rebuttal to the idea that sophistication requires fine, delicate parts. Here, the parts are redundant, distributed, and continuously rearranged.

The redundancy is not only in nuclei. It is also in routes. Because the organism is a network, there are often multiple paths between points. Streaming does not depend on a single tube. If one route is blocked or damaged, flow can reroute through others. The organism's body plan is a moving target, and that makes it resilient in a way that rigid architectures are not. In the forest, this matters when a patch of substrate dries out, when a predator scrapes away a piece, when a competing microbe changes the chemical landscape. In the lab, it matters when a researcher lays down a barrier of salt or light or simply cuts the plasmodium in two. The organism can respond without any one part needing to "know" the whole situation. Local changes in contraction and tube thickness can be enough to reorganize the global pattern.

The plasmodium also invites a more intimate question: what does it *feel* like to be such a thing? Not in the sentimental sense of imagining emotions, but in the sense of sensation and integration. A mammal's body is coordinated by nerves that carry electrical impulses. In *Physarum*, coordination is carried by flows, pressure waves, chemical gradients, and mechanical tensions. The organism senses its world through the same pathways that move its body. Food is not merely detected; it is incorporated into the organism's hydraulics. A region that encounters a rich patch of nutrients can change local metabolic activity, which can alter contraction rhythms, which can alter streaming, which can alter the distribution of resources and signals elsewhere. The "message" is not a discrete packet fired down a wire; it is a shift in the organism's internal economy.

Even the visible pulse—the slight swelling and shrinking of veins, the forward creep of an edge and the gentle recoil—becomes more meaningful when you see it as information-bearing. To watch a plasmodium under time-lapse is to watch decisions embodied in geometry: a tube thicken as if being upgraded from a footpath to a road, a branch dwindle as if a budget has been cut, an entire region stop advancing as if attention has shifted. These metaphors are tempting, and they can mislead, but they also point to something real: in *Physarum*, computation is not separate from construction. The organism does not build a network and then use it; it uses the act of building to determine what the network should be.

The alienness of the plasmodium is therefore not merely a matter of scale. A large cell with many nuclei is already strange, but the deeper strangeness is that the cell's architecture is an ongoing negotiation between physics and physiology. The organism continuously solves the problem of being itself: staying moist, distributing nutrients, avoiding hazards, committing to profitable routes, retracting from wasteful ones. No committee of specialised tissues meets to do this. No central controller issues commands. The cell is the controller, and the controller is a pattern of flows.

Seen this way, the plasmodium can be appreciated as a kind of natural experiment in how far life can push the idea of a cell. It is tempting to treat the textbook cell as the basic unit and multicellularity as the route to complexity. *Physarum* suggests a different possibility: that a single cell can become complex by becoming extensive, internally networked, and rhythmically active. It gains a form of body without becoming many-celled. It gains redundancy without needing organs.

It gains integration without nerves. The price is that it must be always in motion internally, always streaming, always pulsing, because stasis would fragment it into disconnected regions.

It is worth pausing on the aesthetic of this, because popular science often pretends that beauty is a distraction. The plasmodium is beautiful in a way that feels almost improper. Under the right lighting, the veins shine like wet gold. The advancing fronts lace outward with delicate branching patterns that resemble river deltas or lightning. And all of it is alive, not as a metaphor but as a physical fact: the pattern is maintained by living contractions, fed by living transport, remodelled by living growth. When it is starved, it can transform, abandoning the wandering plasmodium for a different architecture entirely, producing fruiting bodies that look like tiny sculpted forests. But when it is fed and active, its beauty is the beauty of a system that is both material and purposeful.

The million-nuclei cell, then, is not a curiosity bolted onto the story of slime-mould intelligence; it is the foundation. The capacity to explore, to connect, to reorganize, to persist after damage, to behave as a coherent whole across centimetres of space—all of it begins with the decision, made by evolution long ago, to keep one membrane and multiply the nuclei within it, and to make the cytoplasm itself into a moving, communicating medium. The plasmodium does not need a brain to coordinate its body because its body already *is* a coordinating device: a living network whose pulse carries both nourishment and news.

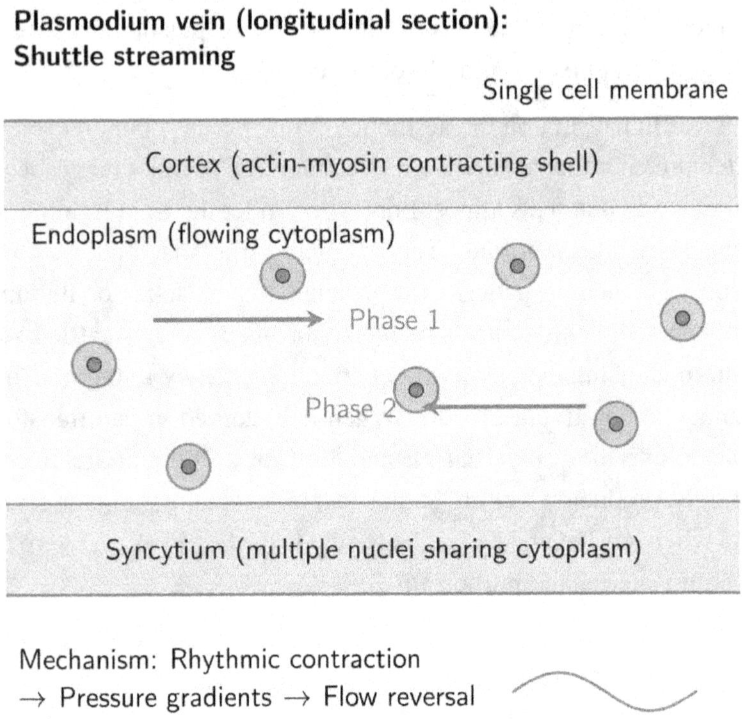

1.2 Thinking Without a Center

Brains make an old promise: if you want coherent behaviour, you need a headquarters. Sensory information comes in, decisions are made somewhere in the middle, commands go out. Even when we acknowledge that nervous systems are complicated webs rather than simple switchboards, we still tend to imagine a privileged place where "it all comes together". The appeal of that picture is not only scientific; it is psychological. A centre is comforting. It offers a location for agency.

A plasmodium of *Physarum polycephalum* refuses to supply such a location. There is no organ you can point to and call the decision-maker. There is not even a stable body plan; the organism redraws its own internal map as it moves. Yet the behaviour can look stubbornly goal-directed: it explores, it commits, it optimizes, it avoids, it sometimes seems to remember. The temptation is to smuggle a brain back in by metaphor—to say that the veins are "like neurons" or that the tubes "compute" in the same way a circuit does. Better to be more literal, and more radical: the plasmodium does not think *despite* lacking a centre; it thinks *by* being a distributed, constantly remade physical network.

The phrase *distributed intelligence* captures this idea in plain terms. Intelligence here does not mean introspection or language. It means the capacity of a system to solve problems that matter for survival: finding food, balancing exploration with exploitation, maintaining connectivity while expanding, adapting to obstacles, resisting damage. Distributed means that no single component contains the full plan. The "answer" is not stored in one place and then executed elsewhere. It emerges from many local interactions: a tube thickening because flow is higher, a branch shrinking because it carries little traffic, a rhythm shifting because nutrients are richer on one side than the other.

A useful way to ground this is to compare network topologies, not as an engineering exercise but as a way to clarify intuition. In a centralized system, many peripheral nodes depend on a hub. In a distributed system, connectivity is spread across many nodes and links. The first makes coordination easy but makes failure catastrophic. The second makes control less straightforward but makes the system robust, because there is no single point whose loss collapses everything.

Centralized control (brain-like hub)

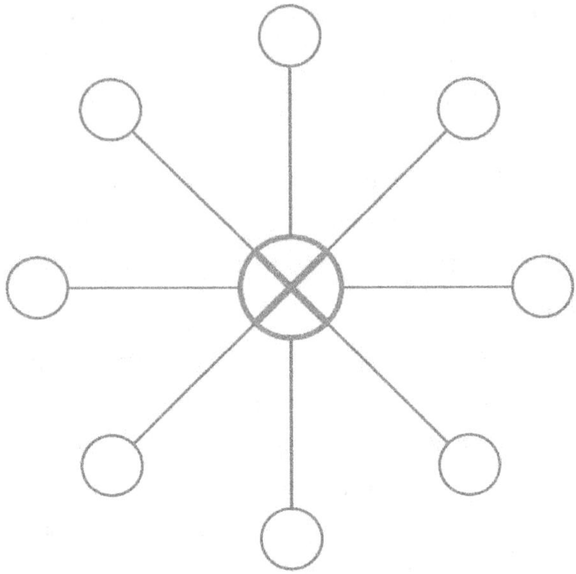

Hub damage disables the network
Fast coordination; single point of failure

Distributed control (slime-mould mesh)

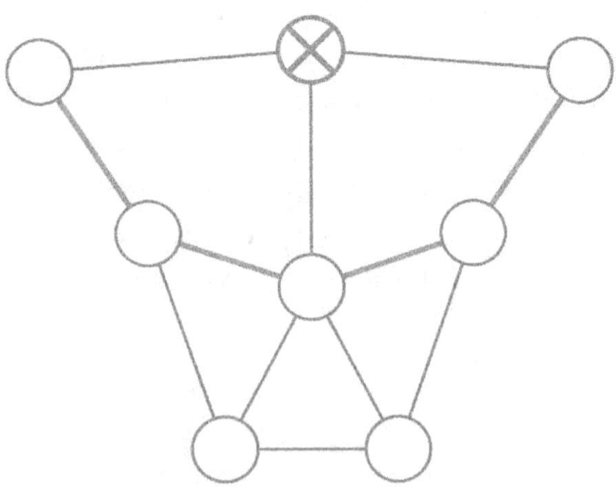

Local loss reroutes flow rather than stopping it

That picture is only a schematic, but it maps onto real biological trade-offs. Brains are extraordinarily good at rapid, integrated responses. They pay for this by being costly tissue, demanding continuous energy and protection. A slime mould has taken a different route. It does not require speed in milliseconds; it requires persistence over hours, and the ability to explore an uncertain environment without committing too early. Its "control" is inseparable from transport: where cytoplasm flows, resources go; where resources go, activity can increase; where activity increases, the physical network is reinforced. The organism becomes a self-editing map of its own history of successes and failures.

The most famous demonstration of this self-editing map entered modern folklore in 2000, when Toshiyuki Nakagaki and colleagues presented a maze experiment that has been repeated in classrooms ever since. Two food sources were placed at different points in a labyrinth, and *Physarum* was introduced. At first the plasmodium spread into many corridors, sending out branches as if it were taking a census of possibilities. Over time, many branches dwindled and disappeared, while a small number of tubes thickened. The striking endpoint was a clean connection between the food sources along a route that, under the conditions of the experiment, corresponded to a minimum-length path.

Part of the charm of that result is how it scrambles our usual hierarchy of explanations. A shortest path is the kind of thing an engineer solves with an algorithm. Here it appeared as a change in anatomy. But the anatomy is also a flow network, and the flow network follows straightforward physical constraints: fluid prefers low-resistance routes; pressure gradients drive motion; thicker tubes carry more flow. Once a tube carries more flow, it becomes even more valuable as a transport route. That is positive feedback: success breeds further success.

Atsushi Tero and collaborators offered a particularly clear mechanistic account in the late 2000s by modelling the plasmodium as a network whose edges change their conductivities in response to flux. In ordinary language, a tube that is used becomes easier to use; a tube that is neglected becomes harder to use and eventually disappears. The rule is local. No part of the network needs to know the whole maze. Yet the outcome can look global and purposeful. It is an example of emergence: a pattern at the level of the whole organism arises from repeated, local adjustments.

Emergence is a word that can become mystical if handled carelessly. It does not mean that the system is doing magic. It means that when many components interact, the combined behaviour can be qualitatively different from what you would expect by inspecting a single component in isolation. A single tube does not "solve a maze". A single local rule about reinforcing flow does not "contain" the idea of a shortest path. But when the organism is constrained by a maze, and when many tubes compete for limited flow, the collective dynamics can lead to a pared-down network that has the properties we would label as an optimized route.

A small but revealing detail is what happens at the start of such experiments. *Physarum* does not choose the shortest path immediately. It explores broadly. The early state is wasteful if you judge it by the endpoint alone, but it is sensible if the environment is unknown. Exploration creates information. In animals, exploration is often framed as a cognitive act: a rat sniffing around a new cage, a person scanning a room. In *Physarum*, exploration is anatomical. The organism makes and tests options by growing into them. Information is written into the pattern of tubes and into the rhythm of flow.

This anatomical information can persist even when the immediate stimulus changes, which is one reason researchers have been tempted to talk about memory. If a region has been a reliable highway for streaming, its tube may remain thick for a while even after the local food is consumed, simply because the reinforced structure does not instantly decay. That is not memory in the sense of a stored representation, but it is a form of embodied history: the organism's present options are shaped by what it has done.

Sometimes the claims go further. In 2008, experiments were reported in which *Physarum* appeared to anticipate periodic events: after being exposed to unfavourable conditions at regular intervals, plasmodia later slowed their movement at the time when the next pulse would have been expected, even when the pulse did not occur. The language of "anticipation" is provocative because it sounds like a creature with an internal clock and an expectation. A more cautious interpretation uses the language of dynamical systems: a periodically perturbed oscillator can become entrained, and the subsequent dynamics can carry echoes of the forcing. The organism is, after all, an oscillatory system: its contractions and streaming already have rhythmic structure. Under repeated periodic stress, the rhythm could be modified in a way that produces timed slowdowns without any need to imagine a mental model of the future.

A similar tension appears in experiments on habituation. Habituation is a simple form of learning: an animal confronted with a harmless repeated stimulus often responds less over time, and the response can recover after a rest. Reports in the 2010s suggested that *Physarum* might show habituation-like behaviour when repeatedly exposed to repellents, becoming less deterred by the stimulus and recovering sensitivity later. If taken at face value, this blurs a boundary many people are attached to: learning without neurons. If taken as physiology, it still blurs a boundary, just a different one: adaptation unfolding on timescales that are fast enough to resemble learning.

The boundary matters because it is easy to say "it learned" and stop thinking. The more interesting question is *how* a distributed body could change its responses based on history. In *Physarum*, several substrates for such change are plausible without invoking anything like synapses. Tube diameters change and then persist for hours. The chemical environment of the cytoplasm changes with metabolism. Contraction rhythms can be modulated by local cues and can spread through the network as waves. Even a simple shift in oscillatory phase relationships across the body can reconfigure where pressure builds and therefore where streaming prefers to go. In other words, the organism has many ways to carry state forward in time, because it is literally a moving medium with internal dynamics.

This is where comparison with brains becomes clarifying rather than dismissive. Animal brains separate signalling and transport. Neurons send electrical signals; blood vessels deliver fuel. The separation allows speed and specificity: a neuron can fire without having to remodel a blood vessel on the spot. But separation also creates dependencies: if the vascular system fails, the neurons fail; if a specific hub fails, downstream regions can lose function. In *Physarum*, signalling and transport are entangled. Flow carries nutrients, but it also carries molecules that affect contraction; contraction shapes flow; flow shapes the network; the network shapes future flow. It is slower, but it is deeply integrated. It is also harder to "break" in the way that centralized systems are broken, because there is no single component that can be destroyed to stop coordination outright.

The famous Tokyo rail analogy, presented by Tero and colleagues in 2010, pushed this point into public imagination. In a stylized experiment, oat flakes were arranged to represent urban centres around Tokyo, and the slime mould was allowed to grow and connect them. The resulting network was compared with the actual rail system, not as a claim that engineers learned from slime moulds, but as a demonstration that similar trade-offs can be reached by very different processes. Transport networks must balance cost (building and maintaining lines), efficiency (short routes), and robustness (alternative routes when one fails). The slime mould, through local reinforcement and pruning, can generate networks that sit in a similar compromise space.

That comparison can be overplayed. Rail planners contend with politics, property, geography, and history. Slime moulds contend with humidity, nutrients, and toxins. Yet the analogy earns its keep by making a general point about distributed intelligence: optimization is not always a top-down calculation. Sometimes it is the stable endpoint of feedback loops acting under constraints. A city does not need a single mind that computes the perfect network from scratch; it can accumulate improvements, reinforce successful routes, abandon failures, and end up with something that looks designed. The slime mould offers a compressed, visible, experimentally manipulable version of that process.

There is another way to express the same idea, one that is closer to everyday experience. If you pour water onto a flat surface, it does not need to "decide" where to go; it follows gravity and the contours of the surface. If you keep pouring, channels can form and deepen, and the flow will increasingly prefer them. Over time the pattern can look purposeful: a drainage network. The difference with *Physarum* is that the surface it flows over is not the only thing changing. The organism changes its own channels actively, under the influence of what the flow is carrying. It is as if the water were also a construction crew, laying pipes where the water tends to go, dismantling pipes that are underused, and shifting the slope of the land in response to the taste of dissolved nutrients.

Thinking without a centre does not mean thinking without structure. The plasmodium's structure is just not the kind we are used to privileging. It is a structure that can be redrafted, and whose redrafting is the means by which the organism makes choices. The "computation" is not in a special tissue; it is in the interplay between geometry and flux.

This perspective also changes how we think about error. Centralized systems can fail dramatically if a key component is damaged or overloaded. Distributed systems fail differently: they can degrade gracefully, becoming less efficient rather than collapsing. A plasmodium cut in half can, under the right conditions, continue as two viable fragments. A portion stressed by light or salt can withdraw while other parts continue feeding. Even within an intact organism, parts can be temporarily sacrificed. This is not altruism; it is a consequence of architecture. When control is spread out, the system can afford local losses. That resilience is itself a kind of intelligence, if intelligence is understood as problem-solving under uncertainty.

There is still a paradox, and it remains deliciously unsettling. A plasmodium looks like it is making a unified choice: *go there, not there*. Yet there is no executive. What exists instead is competition among possibilities, mediated by flows. Branches are like hypotheses written in protoplasm. They are tested by whether they bring back resources and whether they can be integrated into the existing transport economy. The winning hypotheses become thicker tubes; the losers become ghostly threads and then vanish. The organism's "mind" is a shifting pattern of reinforcement.

For humans, the most instructive aspect may be the humility it demands. It is common to treat brains as the pinnacle of problem-solving and everything else as a lesser imitation. *Physarum* suggests that the space of possible intelligences is broader, and that some of the most reliable forms of intelligence may be the least like ours. If you want speed, you build a brain. If you want robustness in a messy world, you can build a body that thinks with its whole substance.

The plasmodium, pulsing in darkness, does not offer its reasons. It offers outcomes: a thickened tube here, a withdrawn fan there, a network that becomes more economical as time passes. When it "chooses" a route between foods, nothing announces the decision; the organism simply stops investing in alternatives. The choice is not a moment. It is a gradual redistribution of attention, where attention is measured in cytoplasm and tension, in the slow accounting of flow.

1.3 From Forest Floor to Petri Dish

The easiest place to meet *Physarum* is not a laboratory but a log. Turn over damp bark in a shaded wood, lift a handful of leaf litter that smells of mushrooms and soil, and there is a chance of finding a yellow web pressed into the grain like spilled paint that has decided to crawl. The forest floor is full of organisms that hide by being small. *Physarum* hides by being transient. One day a patch of decaying wood looks ordinary; the next it is traced with bright veins; a day later it may have withdrawn, leaving only a faint sheen and the suggestion that something happened there.

That "something" baffled naturalists for a long time because it was hard to decide what sort of thing it was. When early microscopists catalogued the living world, they relied on visible structure and life habits: plants rooted and photosynthetic, animals mobile and ingesting, fungi immobile but growing and fruiting. True slime moulds sat awkwardly across these expectations. The plasmodium moved like an amoeba but eventually produced sporangia—spore-bearing structures—that looked fungal. Many were first described by botanists, then argued over by botanists, and stored in herbaria as dried specimens that preserved the fruiting bodies but not the living, creeping phase that made them so peculiar. It is hard to classify a life cycle when the most decisive part behaves like a slow animal and then, in starvation, becomes a delicate architectural object.

The friction between appearance and ancestry became a productive irritation. Taxonomy is sometimes portrayed as mere naming, but naming is a way of deciding which similarities matter. The history of *Physarum* includes repeated shifts in which similarities were treated as deep or superficial. For a period it was natural to treat slime moulds as fungi with odd habits. Later, with improved microscopy and a better grasp of protist diversity, the amoeboid character of the life cycle stages became harder to ignore. Modern biology places *Physarum* among the amoebozoans, closer in broad terms to other amoeboid lineages than to fungi. Yet the old ambiguity left a gift: it trained generations of observers to take seriously the possibility that "fungus-like" and "animal-like" were not reliable guides to the underlying organization of a cell.

Long before the organism became an emblem of biological computation, it quietly became a favourite tool for cell biologists. The reason was not glamour but convenience. Many model organisms are chosen because they are easy to keep alive and easy to manipulate. *Physarum polycephalum* is unusually cooperative in both respects. The plasmodium is large enough to see without a microscope and robust enough to survive being cut and handled. It can be coaxed into dormancy and revived. It grows in the dark, at room-like temperatures, on simple substrates. It will accept oat flakes with an enthusiasm that has made breakfast cereal an unlikely staple of laboratory work.

Most importantly, the plasmodium is a spectacular compromise between the single cell and the multicellular body. It has the experimental tractability of a cell and the physical scale of a small animal. Researchers could extract large amounts of cytoplasm, observe rhythmic contractions, and study transport phenomena without needing to infer everything from fluorescence in a tiny volume. The many nuclei, dividing in synchrony under the right conditions, made

Physarum particularly attractive for studying the cell cycle: when a culture is synchronized, the messy variability of individual cells is replaced by a chorus. Questions that are difficult to address in a mixed population become clearer when millions of nuclei move through the same phases together. This did not yet have the romance of "intelligence," but it established the organism as something more than a curiosity from the woods. It was a system with reliable physiology, reproducible behaviour, and a body plan that could be treated as a physical object.

A subtle change accompanied this move into the lab: the organism began to be understood as an engineered environment. In the forest, *Physarum* encounters gradients and obstacles it did not choose: patches of bacteria, dry zones, bitter chemicals, the irregular geometry of bark. In a Petri dish, the experimenter chooses these features. Agar becomes a standardized landscape. Nutrients become discrete targets. Repellents become controllable boundaries. The lab did not merely observe *Physarum*; it created worlds in which its distributed physiology could be tested.

This shift is so ordinary in experimental biology that it is easy to miss how philosophically loaded it is. A brain is often studied by recording from neurons while the animal is presented with controlled stimuli. For *Physarum*, the "stimuli" are geography itself: food is placed at a coordinate, light is directed at a region, salt is painted as a line. The organism responds by building and dismantling parts of its own body. The output is not a spike train or a muscle twitch but a changing network that can be photographed, measured, and compared across trials. The Petri dish turns physiology into cartography.

The famous maze experiments did not appear out of nowhere; they were a natural consequence of having an organism that expresses its internal decision-making as tube architecture. When Toshiyuki Nakagaki and colleagues reported in 2000 that *Physarum* could reorganize its network in a labyrinth to connect food sources along a minimal route, the experiment felt both playful and unsettling. Playful, because it used a children's puzzle as a scientific apparatus. Unsettling, because it suggested that the behaviours we associate with planning might be rooted in much simpler principles: local reinforcement, flow dynamics, and the physical costs of maintaining redundant paths.

By the time these experiments reached popular audiences, they were often told as a kind of fairy tale: a "brainless" organism solves a problem meant for minds. The laboratory history is more grounded and, in some ways, more impressive. Maze-like tasks are not an arbitrary challenge; they are a way of probing how an adaptive transport network behaves under constraints. A plasmodium is, in its daily life, a problem of routing. It must deliver nutrients from profitable foraging fronts to other regions, and it must do so efficiently enough to survive when food is scarce. The maze simply forces the issue into a shape that human observers can recognise.

A different line of experimentation pushed the story from puzzles toward infrastructure. In 2010, Atsushi Tero and colleagues arranged food sources in a pattern inspired by the region around Tokyo and compared the resulting slime-mould network to the existing rail system. The comparison did not claim that the organism "knew" Tokyo. It revealed something more general: when a network grows under competing pressures of efficiency, cost, and robustness, it can converge on designs that resemble those produced by engineering. The shock of recognition—a biological network resembling a transportation map—was a way of making the abstraction of distributed optimization visible.

This was also the moment when *Physarum* began to be welcomed into computer science and physics departments not as a biological specimen but as an existence proof. Engineers and theorists have long studied how to build systems that work without central control: the internet, power grids, ant-colony algorithms, swarm robotics. The slime mould offered a living example of such a system. It did not solve problems by manipulating symbols; it solved them by pushing fluid through a deformable network and letting local feedback reshape that network over time. For researchers who wanted to understand computation as a physical process rather than an abstract one, *Physarum* was irresistible.

The phrase "biological computing" can mislead if it conjures images of slime moulds replacing silicon chips. The more serious lesson is about what counts as computation in the first place. A Petri dish experiment with oats is not a practical route to faster laptops. It is a demonstration that optimization can occur in matter itself, when matter is organized in a way that couples sensing, transport, and structural change. The plasmodium does not carry out a calculation and then act; the act is the calculation.

That perspective depends, crucially, on the disciplined simplicity of laboratory cultivation. The daily work of keeping *Physarum* alive is prosaic, and that prosaic quality is part of the scientific story. Cultures are often maintained in dark, humid conditions at moderate temperatures, with a regular supply of food. Dormant forms can be revived by rehydration. On agar, an oat flake is both a meal and a beacon: it releases attractive cues, it provides a localized reward, and it becomes a node around which the organism's transport network can reorganize. These routines matter because they turn an organism that appears and disappears in the forest into a stable experimental partner.

It is hard to overstate what that stability enables. In the wild, you might observe *Physarum* for a while and then lose it. In the lab, you can repeat a maze setup dozens of times, vary a single parameter, and ask what changes. You can compare strains. You can cut and rejoin fragments. You can place repellents as barriers or gradients. You can record the oscillations of contraction, quantify the thickness of tubes, and relate geometry to flow. The Petri dish does not tame the organism in the sense of making it behave like something else; it amplifies the parts of its behaviour that are otherwise fleeting.

There is also a cultural history worth noticing. The popularity of slime mould experiments in public demonstrations and classrooms reflects a hunger for intelligences that are not human-shaped. Watching *Physarum* build a network feels like watching thought externalized, made visible. A brain's computations are hidden inside a skull. A slime mould's "reasoning," if one dares to use that word, is spread out on a plate where it can be photographed. It gives the spectator a role that is normally denied in neuroscience: you can see the algorithm taking form.

At the same time, the organism has been a corrective to human vanity. It is tempting to treat the laboratory as the place where the organism becomes impressive, but the lab merely makes certain capacities legible. The plasmodium evolved to handle a forest floor that is patchy, competitive, and unreliable. Microbes bloom and crash. Moisture comes and goes. The terrain is uneven. The plasmodium's network is not a party trick; it is a survival strategy tuned to an environment where the best route today might be a dead end tomorrow. In that sense, the most important invention in the story may not be the maze or the Tokyo-inspired layout, but the very idea of treating a living transport network as the primary object of study.

A final anecdote brings the forest and the lab close together. Anyone who has kept *Physarum* cultures learns, sooner or later, that the organism does not respect the boundaries of an experimenter's intentions. Leave a culture too long without attention, and it may wander away from the neat central area of the dish, climbing the walls toward condensation, seeking humidity, tracing the lid with yellow veins.

Starve it, and it may retreat into a dormant state, a hardened form that looks inert until water returns. Feed it generously, and it can spread with exuberance, turning a plate into a bright, branching map. The organism behaves as if it is always asking the same question: where is the world currently favourable, and how can my body be arranged to exploit it?

That question, posed in protoplasm rather than in words, is what carried *Physarum* from the damp anonymity of the forest floor into the spotlight of modern science. The Petri dish did not give the organism new powers; it gave human observers a clean stage on which to watch a distributed body negotiate with its environment, turning rhythm and flow into a plan that exists nowhere except in the network itself.

A brief timeline of *Physarum* research

- **19th c.** Naturalists catalogue slime moulds; classification disputes reflect their mixed habits.
- **mid-20th c.** Laboratories adopt *P. polycephalum* for cell biology: large, tractable plasmodia; nuclear synchrony.
- **late 20th c.** Quantitative physiology expands: oscillations, streaming, and transport studied as whole-body dynamics.
- **2000** Nakagaki's maze experiment: network pruning yields a minimal route between food sources.
- **2007–2010** Tero and collaborators model adaptive networks and compare slime-mould designs with transport systems (e.g. Tokyo).
- **today** A living case study in distributed intelligence: biophysics meets computation and robotics.

The pattern is less a march of ideas than a change of vantage point: from naming a curiosity to measuring a network.

Chapter 2

The Pulse of the Network

Beneath the apparent chaos of yellow veins lies a sophisticated hydraulic machine acting as both circulatory system and brain. By investigating the rhythmic contractions that pump fluid through its body, we reveal how a single cell harnesses simple physics to process information without a single neuron.

2.1 The Architecture of Flow

A slime mold in its plasmodial stage looks, at first glance, like something spilled: a smear of yellow, a lacy fan, a branching puddle that creeps over bark or agar. The surprise is not that it moves, but that it holds together. Touch it with a probe and the whole body answers, not with a twitch of a limb (it has none) but with a subtle tightening and a slow, purposeful redirection of growth. The organism behaves like a unified animal while lacking the anatomical conveniences animals rely on: no nervous system, no muscles arranged in bundles, no circulatory pump, no organs sealed off behind membranes. What makes the plasmodium so compelling is that it is coherent *without* being compartmentalized. It is one cell on a scale that usually belongs to tissues.

The word *syncytium* names this kind of being: a single cell containing many nuclei sharing one continuous interior. In many organisms, syncytia are local and temporary. In a developing fruit fly embryo, nuclei divide in a common cytoplasm before cell boundaries appear. In vertebrate skeletal muscle, many nuclei sit along a long fiber created by cell fusion. Physarum, by contrast, lives as a syncytium that is neither embryonic convenience nor specialized tissue but the whole organism itself. Millions of nuclei inhabit one cytoplasm, and the cytoplasm is not a stagnant gel. It is an engineered fluid environment, threaded into tubes, pressured into motion, and constantly rebuilt by the very currents it conducts.

That combination—a single interior and a plumbing system—creates a conceptual difficulty. When most of us hear *cell*, we imagine a bounded sack: a surface membrane, a relatively uniform interior, and perhaps a few internal structures floating in it. But a plasmodium is closer to a city-wide utility grid than to a sack. Its boundaries are flexible, it has a thickened scaffold at the periphery, and its interior is differentiated into zones that behave as different materials. The architecture is not decorative; it is the cell's way of making itself large without falling apart.

Consider the basic problem: diffusion. For a small cell, diffusion is enough. Oxygen, sugars, salts, and signaling molecules wander from where they are plentiful to where they are scarce. The time for diffusion grows with the square of distance. Double the distance and diffusion takes four times longer; scale up to centimeters and diffusion becomes a kind of starvation by physics. Any organism that wants to span macroscopic distances must solve transport. Animals solve it with circulatory systems and hearts. Plants solve it with xylem and phloem, with negative pressure and sugar-driven flows. Physarum solves it by turning its entire body into a reconfigurable network of veins, so that transport is not a separate system *inside* the organism; it *is* the organism.

A useful way to picture this is to imagine the plasmodium as a living lacework pressed against a surface. The lace consists of tubes that merge and split at junctions, making loops and alternative routes. The pattern looks strangely familiar to anyone who has seen maps of rivers, road networks, or leaf veins. That similarity tempts one to treat the structure as a metaphor for intelligence: the slime mold builds *good* networks. Yet the more interesting claim is more literal. Those tubes are not merely lines of growth; they are the cell's *mechanical skeleton* and *hydraulic circuitry* at the same time. A vein is not a passive straw; it is a contractile, sensing, remodeling organ made from the same continuous cytoplasm as every other vein.

At the scale of a single tube, the material organization is often described as a two-layer arrangement: a gel-like outer cortex and a more fluid interior stream. The older language calls the outer region *ectoplasm* and the inner region *endoplasm*. These words can sound antiquated, but the distinction is physically real. The cortex behaves like an active elastic sleeve: rich in actin filaments and myosin motors, capable of contracting, stiffening, and anchoring. The interior behaves like a viscous solution carrying organelles, metabolites, and nuclei in motion. Between them lies a shifting boundary rather than a fixed wall; cortex can be assembled and disassembled, and the interior can locally thicken into a more gel-like state. The tube is, in other words, not a pipe made of dead material, but a pipe whose walls are made of the same living substance that flows through it.

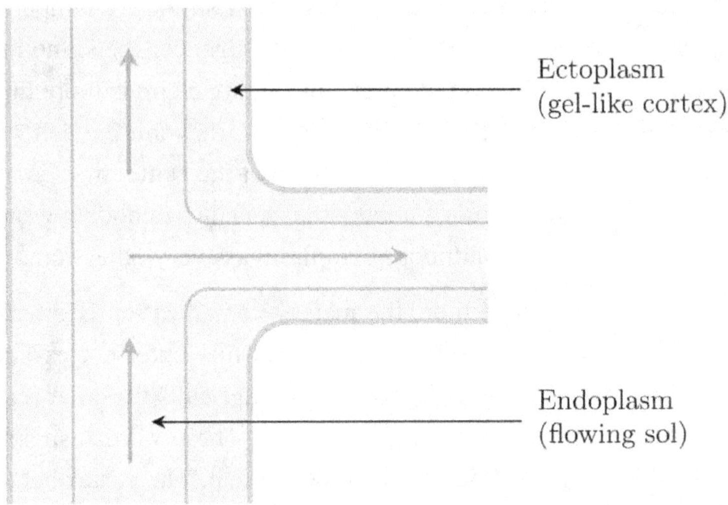

The architectural ingenuity becomes clearer when one asks what a vein must accomplish. It must keep an open lumen for flow. It must resist collapse when local pressure drops. It must be capable of tightening to drive motion. And it must be able to change diameter for longer-term remodeling, widening where traffic is heavy and narrowing where it is not. Engineers would reach for composite materials: a stiff outer layer to bear stress, a flexible inner layer to permit motion. Physarum reaches for something more subtle: an outer layer that is not merely stiff but *active*. The cortex does not just withstand forces; it generates them.

A single tube, however, is only the beginning. The real architectural achievement is that the tubes form a network, and the network spans the whole body. Networks are special because they decouple distance from connectivity. If you want to send nutrients from one end of the organism to the other, you do not need to move through a uniform slab of cytoplasm; you move through a set of conduits that provide low-resistance pathways. Junctions allow rerouting around damage or obstacles. Loops provide redundancy. When the environment changes—a food source appears, a region dries out, a toxin is encountered—the network does not need to be demolished and rebuilt from scratch; it can be tuned by changing tube diameters and by shifting where new growth is reinforced.

An accessible analogy is to think of the plasmodium as a city built entirely of roads, where the roads are also the factories and the citizens, and where traffic flow actively remodels the roads as it happens. That sounds like fantasy, but it captures an important point: in Physarum, structure and function are not separated by design layers. The same physical process that moves material also carries information about what is useful. A tube that repeatedly carries flow becomes a stable feature; a tube that sits idle becomes a candidate for dissolution. The architecture is therefore not only about transport but about *decision-making embodied in matter*.

One concrete historical episode brought this architecture into public view. In 2000, at a time when slime molds were still largely a curiosity outside specialist circles, Toshiyuki Nakagaki and colleagues reported a striking experiment: place food sources in a maze and the plasmodium will, after exploring widely, leave behind a thickened set of tubes tracing an efficient route between the food. Later popular accounts turned this into the claim that slime molds "solve mazes." The headline is catchy but incomplete. The organism does not compute a solution in advance; it physically instantiates many options and then prunes them. The architecture of flow is the medium of this pruning. Exploration produces a sprawling network; sustained flow reinforces certain paths; weak flow allows others to wither. The maze, in this sense, does not become solved *in the slime mold's head* but *in its plumbing*.

To appreciate what is being optimized, it helps to imagine the plasmodium as a sheet of living material that must satisfy competing demands. It needs to cover ground to find resources; it needs to keep resources moving across what it has already covered; it needs to maintain routes that are robust to disturbance; it needs to avoid paying the metabolic cost of maintaining thick tubes everywhere. You can feel the tension between these demands by thinking about a hiking trail system. A dense web of trails makes every destination reachable and provides alternatives when a trail washes out, but it is expensive to maintain. A single main trail is cheap but fragile. Physarum's network lives in this trade-off zone, and the organism's size makes the trade-off unavoidable. A syncytium cannot simply grow larger without also growing a transport system. The network *is* the scaling solution.

The geometry of the network is not random scribble. Photographs of mature networks reveal that junctions frequently have a characteristic three-way branching. This is not a mystical preference for the number three; it is a mechanical and geometric convenience. In a planar sheet-like network, three-way junctions can tile space efficiently and distribute tensions without creating awkward angles that encourage tearing or local collapse. Road engineers know that intersections are expensive and dangerous; too many roads meeting at a point is hard to manage. A three-way junction is, in a sense, the simplest intersection that still allows choices. Again, the language of choice is tempting, but the deeper point is physical: a junction is a point where flows can divide and where stresses can be redirected, and three is often the stable number for a branching in a deformable medium.

Yet the architecture is more than its topology—more than who connects to whom. Two networks can share the same connectivity and behave very differently if their tubes differ in length and diameter. In fluid transport, radius matters dramatically: for laminar flow in a cylindrical tube, resistance scales roughly with the inverse fourth power of radius. Doubling the radius can reduce resistance by a factor of sixteen. Even if the details in living tubes depart from ideal cylinders, the lesson remains: small changes in tube thickness can have outsized effects on where flow prefers to go. Physarum can therefore tune its internal traffic with modest morphological adjustments. A tube need not be eliminated to become less important; it can simply narrow, raising its resistance so that flow is diverted elsewhere.

This sensitivity gives the plasmodium a kind of *analog control* that is hard to appreciate from still images. In a photograph, a tube either appears present or absent. In the organism, tube diameters are continuously graded and dynamically fluctuating. A region that seems "thin" in one moment can thicken over the next hour if it becomes part of an important route. Conversely, a thick tube can gradually deflate if the environment no longer supports its use. The architecture is therefore best thought of as a living map of priorities: a material record of where resources have been, where they are needed, and where the organism expects to need them soon.

This is also why the plasmodium is so hard to divide into parts. If you cut it, each part can survive and regrow, which might suggest modularity. But while intact, it operates as a strongly integrated hydraulic body. Pressure differences and flows couple distant regions. A food source on one side can change contraction patterns and streaming in tubes far away. The network makes the body mechanically and chemically continuous. In animals, long-distance coordination is largely electrical and chemical through nerves and hormones. In Physarum, a great deal of coordination is hydraulic and mechanical, carried by the movement of cytoplasm itself and by the stresses that movement generates in the cortex.

A small anecdote from laboratory life captures this integration. Researchers who culture Physarum on agar learn quickly that the organism is sensitive to the moisture of its world. If the agar begins to dry, the plasmodium often responds not by continuing its explorations but by consolidating: it retracts from thinner fans and pulls mass into thicker, more protected cords. To a casual observer this looks like timidity, as if the organism is "deciding" to hide. But the architecture tells a more grounded story. Thin peripheral sheets have high surface area relative to volume and lose water quickly. Thicker tubes have lower surface-area-to-volume ratios and are mechanically sturdier. Consolidation is not merely a behavioral choice; it is a re-engineering of the transport-and-storage geometry to better fit a drying environment. The network is the organism's way of managing physical risk.

The language of risk points to another architectural feature: redundancy through loops. A tree-like network (no loops) is efficient for distributing material outward from a source, but it is vulnerable: cut one branch and everything downstream is isolated. A loopy network can reroute around damage. The plasmodium often maintains cycles, especially in well-fed, stable conditions, and these cycles can be read as insurance policies. But insurance has a cost. Maintaining an extra thick tube is metabolically expensive. The organism therefore tends to keep many thin alternative connections and a few thick backbones. This resembles the way human infrastructure is built: many small streets feed into a few arterial roads; power grids include redundancy but concentrate capacity in major lines. The resemblance is not because Physarum is imitating cities, but because both are solving constrained transport problems in space.

What gives Physarum its distinctive flavor is that the architecture is not laid down by a developmental blueprint. There is no genetic plan specifying that a tube of a certain width must connect two precise points. Instead, the architecture is produced through continuous feedback between flow and form. When cytoplasm moves through a tube, it exerts shear stress on the inner surface and pressure on the wall. The cortex responds: actomyosin activity can adjust tension; growth machinery can add or remove material; local thickness can increase or decrease. Over time, this creates a tendency for frequently used routes to become reinforced. If you have ever watched a dirt path form across a grassy field, you have seen an externalized version of this principle: repeated use suppresses grass regrowth and compacts soil, making the path more likely to be used again. In Physarum the "foot traffic" is internal flow, and the "soil" is living cytoplasm, but the feedback logic is similar.

This feedback is not only local. Because the plasmodium is a connected network, changing one tube changes the distribution of pressures and flows everywhere. Widen a tube in one place and you reduce resistance there, which can increase flow through that route and reduce flow through competing routes. Those changes, in turn, trigger remodeling elsewhere. The architecture is therefore a system of coupled adjustments rather than a set of independent modifications. This coupling allows the organism to produce coherent global changes from local rules, the essence of distributed intelligence in a body without neurons.

The syncytial nature of the organism also means that nuclei share a common chemical environment, but they are not merely passive passengers. A plasmodium contains many nuclei that are genetically identical (under typical laboratory conditions) and yet the organism can maintain spatial differences in activity. Regions can differ in growth, contractility, and sensitivity. One way to think about this is to imagine a parliament without walls: everyone hears the same general chatter, but local circumstances still matter. A region near a food source experiences higher concentrations of nutrients and perhaps different chemical cues; a region at the exploratory edge experiences different mechanical stresses and contact with the substrate. In such a body, position is information. Architecture supplies the stable channels through which that information can be shared widely without erasing local differences.

Even the term "vein" can mislead if it suggests a fixed set of conduits. In Physarum, veins are provisional. They appear, strengthen, merge, and vanish. Their walls can thicken into robust cords or thin into fragile threads. This plasticity is not a secondary feature; it is the strategy. A nervous system is powerful partly because it allows rapid reconfiguration of behavior without rebuilding anatomy. Physarum, lacking nerves, must reconfigure behavior through changes in physiology and morphology, and its morphology is unusually quick to change because it is built from soft, living material. Architecture and behavior are therefore intertwined. When the organism "chooses" to connect two food sources more directly, it does so by physically sculpting a lower-resistance route between them. The body becomes its own memory of what worked.

Seen this way, the plasmodium is neither a blob nor a miniature animal. It is an extended, hydraulically integrated cell whose size is made possible by a liquid lattice: a network of contractile tubes with a gel-like cortex and a flowing interior, continuously tuned by the demands of transport. The architecture is what allows a single cell to be wide without being thin, connected without being rigid, and responsive without a central controller. If intelligence is the ability to coordinate many local actions toward a coherent outcome, Physarum achieves it by making coordination a property of its material design, written into the very paths along which its cytoplasm runs.

2.2 A Heartbeat Dispersed

Place a healthy plasmodium of *Physarum polycephalum* on a moist surface and watch it closely—not for minutes, but for the length of a song. A hand lens helps, and a microscope helps more, yet even the naked eye can catch the creature's most characteristic motion: a faint, rhythmic swelling and relaxing that passes through its tubes like breath through bellows. The whole body seems to pulse. This is not an ornament of life, like the flutter of a leaf in a breeze; it is the engine. The plasmodium's circulation and its coordination share the same mechanism, a back-and-forth flow called *shuttle streaming*.

"Back-and-forth" sounds wasteful if one is used to arteries and veins pushing blood one way. In a human arm, fluid that moved forward for a minute and then reversed for a minute would feel like failure. In *Physarum*, reversing is the point. When the interior fluid surges forward, it brings oxygen, sugars, and metabolites to regions that need them. When it surges back, it does not simply undo the trip; it mixes, stirs, and spreads signals through the network. Over repeated cycles, the organism gains the advantages of both transport and communication, without installing a central pump. The heartbeat is dispersed: a rhythm made everywhere and felt everywhere.

Shuttle streaming is easiest to picture in a single tube. The tube is lined by an active cortex, dense with actin filaments and myosin motors—the same molecular machinery animals use for muscle contraction. When myosin pulls on actin, the cortex contracts, narrowing the tube. Narrowing raises pressure locally and pushes the more fluid endoplasm away from the contracting region. If the contraction is coordinated along the tube, it behaves like peristalsis, the squeezing wave that moves food through an intestine. But *Physarum* adds an unusual twist: the direction of the wave, and therefore the direction of flow, reverses periodically. Pressure gradients flip. The organism alternates between two directions of pumping, typically on a timescale of minutes.

The older literature sometimes describes this as if the slime mold had a built-in metronome. In reality, the rhythm is not imposed from above. It is an emergent oscillation of an active material: a gel that contracts, relaxes, and contracts again because of internal chemical feedback. A key player is calcium, a small ion with outsized influence in cell biology. In many cells, a transient rise in free calcium triggers contraction. In *Physarum*, the relationship is subtler: modern measurements suggest calcium dynamics can be nearly anti-correlated with tube radius, consistent with a role in inhibiting contractility under some conditions. The details are still debated, but the broader point is robust: chemical changes in the cortex couple to mechanical strain, and the resulting deformation feeds back onto chemistry. This is what an oscillator is made of: a loop.

Historically, the fascination with these rhythms is almost as old as serious laboratory work on slime molds. In the 1970s, K. E. Wohlfarth-Bottermann used ingenious tensiometry to measure contractions in strands of *Physarum*, teasing apart longitudinal and radial components and showing that the periods are on the order of minutes under typical laboratory temperatures. Those careful mechanical measurements mattered because they anchored a biological spectacle to numbers. A plasmodium might look like a simple smear, yet it has a measurable, temperature-sensitive rhythm that can be treated as physiology rather than mystery.

Why should temperature matter? Because contraction is a molecular process. Myosin heads bind, pull, release, and reset, and every step depends on thermal motion and chemical reaction rates. As temperature rises within a tolerable range, the cycle quickens; as it falls, everything slows. Anyone who has watched bread dough rise in a warm kitchen understands the intuition. With *Physarum*, that intuition becomes a way to read the organism: a slower pulse hints at a colder plate, a stressed plasmodium, or a change in the chemical environment. The rhythm is not only a pump but a diagnostic of the organism's internal state.

Even if one accepts that the tubes can generate a rhythm, a harder question remains: how can a rhythm generated locally move fluid across a complicated, branching network? A peristaltic wave in a single tube is straightforward. A peristaltic pattern across a random mesh of tubes seems, at first, like a recipe for stalemate—pushing here while pulling there. Yet experiments and theory in the last couple of decades have made clear that network peristalsis can be organized without a conductor. Karen Alim and collaborators showed that *Physarum* expresses coordinated contractions across many tubes, and that transport through the network is maximized when the characteristic wavelength of the contraction pattern is comparable to the size of the organism. That statement sounds technical, but it is a beautifully simple idea. If the organism is small, a short contraction wavelength can move fluid effectively through it. As it grows, a small-scale pumping pattern would merely churn fluid locally. Efficient long-range transport requires pumping patterns that span the body.

This scaling requirement reveals something often missed in popular accounts. The plasmodium is not simply a network with a rhythm laid on top. The rhythm adapts with size. A larger individual tends to develop longer-range coordination in its contractions, not because it is planning ahead, but because transport demands select for contraction patterns that actually move material. In an active, remodeling organism, patterns that fail to move nutrients will starve the regions that produce them. Patterns that succeed will support growth and therefore persist. The heartbeat is not fixed; it is tuned by the physical consequences of beating.

Shuttle streaming also changes how we think about what "signal" means in such a creature. In animals, signals are often electrical spikes or hormone pulses that travel on specialized pathways. In *Physarum*, the moving cytoplasm itself becomes a carrier of information. Suppose a patch of the plasmodium touches food. The local chemistry changes: nutrients enter, metabolites accumulate, and signaling molecules associated with feeding and growth rise. As streaming proceeds, those molecules are carried through the tubes, mixed, diluted, and delivered to distant regions. This is not a perfect broadcast—concentrations decay and are shaped by flows—but it is a genuine communication channel. The organism does not need neurons to let one region influence another; it needs circulation.

Circulation does not merely transport chemicals; it transmits *mechanical* information. A contracting tube changes pressures elsewhere, and pressure changes can modulate contractility. In that way, the network behaves like a coupled system of oscillators. If one region begins to contract more strongly, it can entrain nearby regions, pulling their rhythms into phase. Under certain conditions, waves of contraction sweep across the plasmodium, coordinating the flow through many branches. Under other conditions, the pattern is more patchy, as if different neighborhoods are beating to slightly different drums. Both regimes can be useful. A strongly coherent wave can move material long distances. A more disordered pattern can mix locally and keep the organism responsive rather than locked into a single mode.

To make the back-and-forth idea concrete, imagine a simple stretch of tube between two junctions. During the forward phase, endoplasm streams to the right with a high velocity; organelles and nuclei are carried along like debris in a river. The tube wall may thicken slightly as cortex contracts in a wave. Then comes a brief pause: flow slows, pressures equilibrate, the tube relaxes. Now the reverse phase begins: endoplasm streams left. If this reverse were perfectly symmetric, nothing would be gained; a particle would return exactly where it started. But symmetry is rare in biology. The reverse phase can be slower, shorter in duration, or routed through slightly different branches because the network is not a single straight pipe. Over a full cycle, the result can be a net bias: a subtle transport in one direction, or a preferential delivery of fluid to one region because of how resistances and junctions distribute flow.

This is where shuttle streaming stops being a curiosity and becomes a mechanism for action. A plasmodium often crawls as a whole, extending at its front and retracting at its rear. How can it make net progress if its internal flow reverses? One answer is that the oscillations are coupled to shape changes. Measurements of streaming velocity fields and body deformation have shown that forward and backward phases need not be equal in their effectiveness. A forward surge may push fluid into a broad, advancing fan where new cortex is assembled and adhesion to the surface is strong. The backward surge may pull from a narrower, retreating region where tubes are consolidating. The same back-and-forth flow can therefore have different consequences at different ends of the body, producing net displacement over time. The plasmodium does not need a one-way pump; it needs a biased cycle.

A visual helps here because the logic is temporal. The organism's circulation is not a static map but a repeating pattern of phases. The following diagram compresses a two-minute cycle into four snapshots: forward flow, a pause, backward flow that is shorter or weaker, and the resulting net gain.

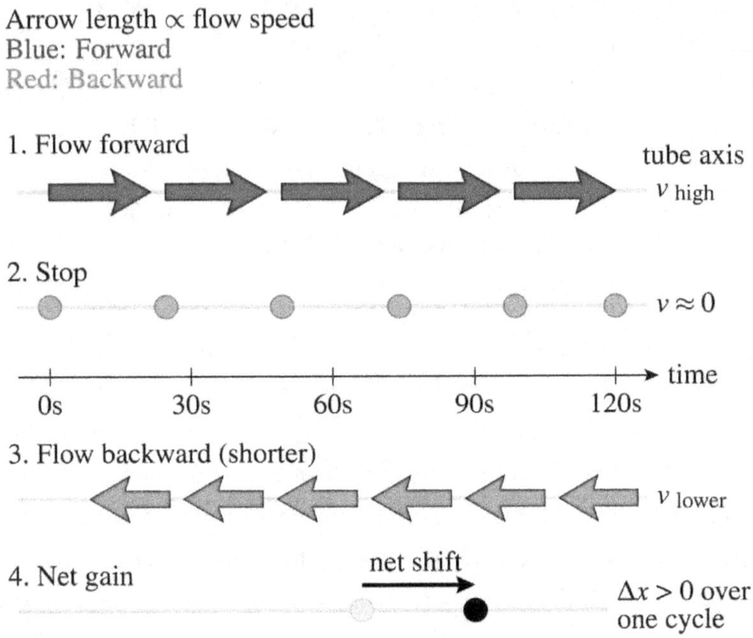

The diagram is deliberately schematic, but the key idea is realistic: oscillation does not imply zero progress. Engineers exploit the same principle in contexts that would otherwise be paradoxical. If you jiggle a box with asymmetric friction, it creeps. If you vibrate grains in a ratchet-like container, they drift. In low Reynolds number fluids—the regime of tiny swimmers—reciprocal motion often fails to produce locomotion, yet clever asymmetries and shape changes can break the symmetry. *Physarum* lives in a world where viscous forces matter, and it has abundant opportunities for asymmetry: branching tubes, changing diameters, different anchoring to the substrate, and unequal timing across the network.

A concrete laboratory example makes the link between streaming and behavior feel less abstract. Offer the plasmodium a choice between two oat flakes placed at different distances. Typically it spreads a fan of thin veins outward, sampling both possibilities. Once one route is established and feeding begins, streaming patterns in the thicker connecting tubes become more pronounced. The fed region often shows stronger contractions, and the network reorganizes so that the main tubes between food sites thicken. The striking point is not merely that nutrients are delivered; it is that the circulation carries a message: *this route matters.* The organism's decisions are written into the amplitude and routing of flow. If the oat flake is removed, the pattern changes again; flow through now-useless tubes declines, and the organism reallocates its pumping to other regions. Even without measuring molecules, one can watch the physiology of attention shifting across the body.

At a deeper level, shuttle streaming helps resolve a fundamental tension between two kinds of transport: *delivery* and *mixing*. A purely one-way river delivers material downstream efficiently, but it can leave side branches stagnant. A purely oscillatory flow can mix well, but seems poor at getting anything from A to B. *Physarum* uses oscillation to do both. The back-and-forth motion repeatedly flushes side tubes, preventing dead zones. It also enhances dispersion beyond what diffusion alone could accomplish. Each reversal stretches and folds concentration gradients, the same way stirring cream into coffee makes it spread faster than waiting for diffusion. Over many cycles, molecules that begin localized can become broadly distributed, and local signals can become whole-body states.

This mixing has emotional weight when one remembers what the plasmodium is trying to do: survive in a patchy world. On a forest floor, food is not a uniform soup. It is discrete: a bacterium-rich patch here, a decaying leaf there, a fallen seed elsewhere. A large, spread-out plasmodium risks becoming a collection of starving fringes unless it can redistribute the gains from a lucky find. Shuttle streaming is the organism's way of making fortune portable. A region that finds food can, through repeated pulses, subsidize distant regions that are still searching. In a multicellular organism, this is the role of circulation. In *Physarum*, it is achieved without a heart, because every tube wall is a little bit heartlike.

The dispersed heartbeat also shapes the organism's sense of itself. A body that shares a circulating interior can coordinate growth and contraction over distances that would otherwise fragment it into semi-independent patches. A chemical cue encountered at one margin can be felt elsewhere not because a signal travels down a dedicated wire, but because the entire interior is in motion, repeatedly sampling and sharing local conditions. The plasmodium becomes, in a literal physical sense, a network that thinks with its blood.

In the dim light of a culture dish, the pulse can seem gentle, almost soothing. Yet it is relentless. Minute after minute, hour after hour, the organism keeps squeezing and releasing, driving currents through its ever-changing lattice. The rhythm is slow enough to watch and fast enough to matter, a reminder that intelligence does not require speed so much as persistence and coherence. A slime mold keeps time with its own body, and that time carries food, stress, and opportunity through every branch until the whole network tilts toward what sustains it.

2.3 The Algorithm of Resistance

A pulsing network that moves fluid back and forth faces a quiet tyranny: friction. Every time cytoplasm is pushed through a tube, energy is lost to viscosity, the internal "stickiness" of fluid. A thick tube wastes less energy than a thin one, but it costs more to build and maintain. A dense web of alternative routes gives resilience, but it also creates miles of plumbing to keep pressurized, repaired, and contractile. The plasmodium lives inside this trade-off, and its most distinctive talent is not merely that it forms networks, but that it keeps editing them. Some tubes swell and become highways. Others fade, collapse, and vanish. The edits are not arbitrary. They follow a rule so simple it sounds almost moral: *use it or lose it.*

The phrase is a convenient shortcut for a physical principle. Flow through a tube generates shear stress along the wall: drag that tugs at the cortex. Flow also generates pressure differences that stretch the tube and change the tension in its active gel. Those mechanical signals are local and immediate. The cortex does not need a map of the entire organism to respond; it only needs to "feel" whether it is being worked. Over time, the organism invests where mechanical work is concentrated, and divests where it is not. The result is a network that tends to reduce resistance along important routes, pruning inefficient loops while keeping just enough redundancy to avoid catastrophe when conditions change.

It helps to make resistance concrete. If you sip a thick milkshake through a narrow straw, your cheeks do the work of overcoming viscous losses. Switch to a wider straw and the drink comes easily. The straw's radius matters more than intuition suggests. In idealized laminar flow, resistance scales roughly with the inverse fourth power of radius. Halve the radius and you do not double the effort; you multiply it by sixteen. Real slime mold tubes are not perfect cylinders and the fluid is crowded with organelles, but the qualitative lesson remains: small changes in diameter reshape the entire economics of transport. A tube that thins slightly can become effectively irrelevant. A tube that thickens slightly can become a dominant route.

That steep dependence creates an opportunity for a distributed algorithm. An algorithm, in this broad sense, is a rule that converts local information into a global outcome. Here the local information is shear and pressure; the global outcome is a network that approximates efficient transport between multiple sites. A single tube "decides" nothing, yet by reacting to its own mechanical burden it participates in a collective computation. The computation is not performed in symbols; it is performed in geometry.

One can see the rule in action in a familiar laboratory tableau. A plasmodium placed on an agar plate with scattered oat flakes will often spread into a broad fan, laying down many thin tubes. This exploratory phase is exuberant and redundant, as if the organism is unwilling to bet too early on any one route. As soon as feeding becomes established, the pattern tightens. Thick cords form between the most productive sites; thinner detours are gradually abandoned. If a food source is removed or a new one appears, the pattern shifts again. The edits are slow enough to watch over hours, but fast enough to matter to an organism that lives by finding scattered patches of nutrition.

The fascination of these behaviors reached a wide audience after a different kind of demonstration. Atsushi Tero and collaborators (2010) compared networks formed by *Physarum* connecting oat flakes arranged like major urban centers around Tokyo with the layout of the Tokyo rail system. The slime mold's network was not a perfect match, nor should it be. Yet it achieved a similar balance among competing criteria: relatively short travel distances, reasonable construction cost (in the biological sense of total tube volume), and tolerance to failures. The point was never that engineers should outsource design to petri dishes. The point was that a physical organism, following local rules, can land in the same design space that human planners occupy when they must trade efficiency against robustness.

For readers wary of anthropomorphism, it is worth translating this into the language of constraints. If you want to connect multiple resource sites, you can minimize total length of tubes, or minimize travel distance between sites, or maximize fault tolerance. You cannot optimize all of these at once. Any real network is a compromise. What makes *Physarum* remarkable is that its compromise is not frozen. It is continually renegotiated by the flows running through it.

A mathematical formulation of the "use it or lose it" rule made this idea sharper and more portable. Around 2007, models of *Physarum*-like adaptive networks were proposed in which each edge of a network has a conductance (how easily it carries flow) that increases when flux is high and decreases when flux is low. The equations look, on paper, like plumbing with a memory. Feed the model two sources and it tends to converge toward a set of high-conductance paths that connect them efficiently; in a maze-like graph, it often strengthens a short path and weakens longer alternatives. Such models are not claims that the plasmodium solves differential equations. They are attempts to capture, in a minimal way, what the organism seems to do in matter: reinforce the routes that actually deliver.

The deeper story sits at the branching points, where tubes meet and choices are made by physics. Consider a junction where one tube splits into two. If both branches have equal resistance, flow divides equally. If one branch is slightly wider, it has lower resistance, so it receives more flow. More flow means more shear and perhaps more stretching, which tends to promote further widening. That is positive feedback. Left unchecked, positive feedback would drive the system to an extreme: one branch becomes a dominant highway and the other collapses. Sometimes that is exactly what is needed, as when one route is clearly better. But the plasmodium must avoid brittle overcommitment. Conditions change; a highway that serves today's food may lead to nothing tomorrow.

Two mechanisms keep the positive feedback from becoming pathological. The first is cost. Maintaining a thick tube requires materials and metabolic energy. The cortex must be renewed, actomyosin machinery maintained, and the tube protected against mechanical failure. Thickening everywhere would be ruinous. The second is the nature of the pumping itself. Shuttle streaming reverses, pressures fluctuate, and the network is loopy. Even if one branch is favored on average, the oscillatory flows can keep some traffic in alternatives, delaying total collapse and preserving a degree of redundancy. The organism's heartbeat, in other words, acts as a kind of regularizer: it discourages the network from locking into a single rigid solution too quickly.

These considerations have a cousin in a classic piece of physiology: Murray's law. Cecil Murray, working in the early twentieth century, asked why blood vessels have the radii they do. If you assume that the body is trying to minimize the total cost of moving blood (energy lost to viscous dissipation) plus the cost of maintaining blood volume, you can derive an elegant relationship between the radius of a parent vessel and its two daughters at a branch. In its simplest form, the parent radius cubed equals the sum of the cubes of the daughter radii. Many biological vascular systems approximate this relationship. The law is not perfect, and it does not dictate every detail of living networks, but it offers a powerful intuition: when you balance friction against construction cost, specific geometries become favored.

Physarum is not a mammal with a fixed circulatory system, but it lives under the same basic physics. Its tubes carry viscous fluid; its body must pay for tube volume. If its network remodeling tends to reduce wasted dissipation while keeping total tube material in check, it is not surprising that some of its branch geometries resemble those of animal blood vessels. The resemblance does not require evolutionary convergence in a strict genetic sense. It can be a convergence in solutions imposed by the mathematics of fluid flow.

The appeal of the resistance algorithm becomes even clearer when one considers failure. A tree network is efficient but fragile. A loopy network is robust but costly. The plasmodium often maintains a backbone of thick tubes supplemented by a mesh of thinner loops. The backbone carries most of the flux, like a highway system; the loops provide alternate routes, like side streets and local detours. If a thick tube is cut, nearby loops can temporarily compensate, keeping regions supplied until remodeling produces a new backbone. This capacity is not heroic; it is practical. In a natural habitat, tubes are damaged by drying, abrasion, and predators. An organism that cannot cope with cuts will be routinely dismembered. The algorithm of resistance thus has a conservative aspect: it avoids both extremes, neither proliferating costly redundancy nor pruning so aggressively that the network becomes a single point of failure.

The same rule also creates something that looks uncannily like memory. Suppose a plasmodium has repeatedly traveled a particular corridor on a plate. The corridor's tubes tend to be thicker because they have carried flow. If the organism retracts and later returns, those thicker remnants can bias the new flows, making it easier to re-establish the same route. The past leaves a physical trace: altered conductance. Unlike a nervous memory, which is stored in synaptic strengths inside a fixed anatomy, this memory is stored in the anatomy itself. It is a memory written in resistance.

That idea can be tested in a simple kitchen-table way. Hobbyists who culture *Physarum* sometimes lay down paper bridges or grooves on agar. After a while, the slime mold tends to favor the established paths, because they have become low-resistance conduits and, often, because they retain moisture. The organism looks as if it "prefers" a route. The preference is not psychological; it is an embodied consequence of past flow and past construction. Anyone who has walked the same shortcut across a lawn and watched it become a dirt path has participated in a similar feedback, except that the slime mold does the walking and the landscaping simultaneously.

Remodeling changes not only which places receive nutrients, but how quickly signals spread. When a network is dense and full of loops, oscillatory flow can slosh around in many directions. That promotes local mixing but can slow the time it takes for a chemical change in one region to influence the whole body, because the flow is distributed across many routes. When the network coarsens into fewer, thicker tubes, long-distance transport can become more efficient. There is also a subtle effect that physicists call Taylor dispersion: in a tube with shear flow, a dissolved substance spreads faster along the tube than it would by diffusion alone, because fluid in the center moves faster than fluid near the walls, and diffusion between these layers converts that shear into enhanced longitudinal spreading. Thick tubes with substantial flow can therefore act as rapid communication lines, not merely as pipelines for calories.

Here the algorithm of resistance becomes an algorithm of attention. By thickening some routes, the plasmodium creates channels that quickly carry the chemical consequences of a local discovery to distant regions. Those channels allow the organism to coordinate growth and contraction across its span. By thinning other routes, it reduces noisy circulation through irrelevant areas, focusing pumping power where it has return. In a nervous system, attention is often imagined as a filtering of signals; in *Physarum*, attention can be achieved by filtering *flows*.

One might ask what, physically, tells a tube to widen or shrink. The simplest story is shear: high shear stimulates growth, low shear permits decay. But living tubes are not passive. The cortex is an active, anisotropic gel, able to generate tension and respond nonlinearly to stress. Pressure inside the tube matters as much as shear along the wall. Recent theoretical work has argued that the coupling between hydrostatic pressure and the mechanics of the actomyosin cortex could set a kind of target shear rate for stability, so that tubes adjust until mechanical stresses sit in a preferred range. Experiments increasingly suggest that calcium dynamics, contractile activity, and radius changes are tightly linked, though the precise timing relationships are still being refined. The honest picture is that the plasmodium's remodeling rule is not a single lever labeled "shear." It is a web of mechanochemical feedbacks that, when coarse-grained, looks like shear-driven adaptation.

This is where the "algorithm" metaphor earns its keep without becoming magical. A computer algorithm is defined at an abstract level: input, rule, output. Many different machines can implement the same algorithm. *Physarum* implements its resistance algorithm with actin, myosin, calcium, membranes, and water. It does not compute shortest paths because it has a concept of shortestness. It tends toward efficient paths because tubes carrying sustained flux become easier for flux to traverse, and because material is limited. The global behavior is the shadow of local physics.

There is also a moral dimension, if one wants it, and it is not the usual one about "nature being efficient." The plasmodium is efficient only when efficiency serves survival. In a stable, resource-rich environment, it can afford redundancy, and it often keeps more loops. Under stress, it consolidates, sometimes retracting into a few thick cords. Efficiency and resilience are not virtues but tactics, selected by circumstance. A human city that optimized only for efficiency would be brittle; one that optimized only for redundancy would be impossibly expensive. *Physarum* navigates the same compromise, guided not by committees but by resistance.

In the end, the most striking feature of the slime mold's network is that its logic is written in a language older than neurons: pressure, viscosity, tension, and chemical feedback. A tube that carries the organism's lifeblood becomes, through that very act, more capable of carrying it again. A tube that falls out of use becomes thinner, quieter, and finally disappears. The organism continually edits its own body to reduce the wasted effort of pumping and to preserve the routes that matter, until the network itself becomes a living record of where energy has been spent wisely.

Chapter 3

Escaping the Labyrinth

Imagine a creature with no eyes to see a maze and no brain to map it, yet it finds the quickest exit with mathematical precision. We venture into the wet lab to witness the slime mold's most famous feat, uncovering how a simple rule of fluid mechanics creates the illusion of sophisticated foresight.

3.1 The Geometry of Hunger

A maze looks like a puzzle because it is drawn for eyes. For a slime mould, the walls and corridors are less like a riddle and more like a landscape that constrains where a body can spread. *Physarum polycephalum* is not a creature with a head that chooses a direction. It is a single, continuous living mass—a plasmodium—that can be stretched into a thin, shimmering sheet or gathered into rope-like veins. When it solves anything, it does so by growing, feeling, and remodeling itself. The word for its most reliable urge is an old one: *chemotaxis*, meaning movement oriented by chemicals. Hunger, here, is geometry: gradients in space that can be climbed.

Chemotaxis begins outside the organism. A food source is not merely a lump of nutrition; it is a local factory of molecules. Oat flakes, the laboratory standby for coaxing *Physarum*, leak a rich mixture of soluble compounds into moist agar. These molecules diffuse outward, more concentrated near the oat, thinner farther away. Diffusion turns appetite into a field: if you could plot concentration in three dimensions, you would get a soft hill centered on the food. The slime mould does not need to know where the hilltop is. It only needs the modest ability to compare conditions here with conditions a millimetre away.

That comparison is trickier than it sounds for an organism without a front end. A plasmodium is a continuous body; any point along its edge can become the leading edge. The word *sensing* can mislead, as if there were a sensor somewhere. Instead, the surface of the slime mould is a responsive skin. When attractant molecules bind to receptors in the cell membrane, they set off biochemical cascades inside the cytoplasm. Those cascades do not merely whisper instructions; they change the physical rhythm of the organism.

One of the most striking features of *Physarum* is that it pulses. Under a microscope, the cytoplasm streams back and forth in a slow, tidal motion, driven by rhythmic contractions of an actin-rich cortex. This oscillation does two things at once. It transports materials through the body, and it creates a built-in timing mechanism that can be modulated. In the presence of a stronger attractant signal, the local contraction rhythm can speed up or increase in amplitude. The organism effectively turns chemistry into hydraulics: a faint external difference becomes a local change in pumping.

Imagine the edge of the plasmodium as a coastline. Along that coastline, some stretches sit in slightly richer chemical water than others. Where the chemical signal is stronger, contractions strengthen; where it is weaker, they slacken. A strengthening contraction does not tug the organism like a muscle pulling a limb. It changes internal pressure and flow. Cytoplasm is pushed more insistently through some tubes than others, and the body thickens where flow and delivery are more generous. Growth is not a decision laid on top of the body; growth is the body expressing its physics under biased conditions.

The same logic applies to avoidance, which is as important to maze behaviour as attraction. *Physarum* dislikes bright light and desiccation; it is happiest in dim, damp places that resemble the underside of leaf litter. Light can function as a repellent: areas exposed to illumination become less hospitable, and the organism retracts or refuses to advance into them. In a maze experiment, the walls constrain movement, but repellents can do something subtler: they sculpt where the organism is willing to invest its thickness. When a corridor is slightly more stressful, it becomes a less profitable path, even if food lies beyond. In human terms, the slime mould does not merely chase a smell; it is constantly balancing temptation against discomfort.

This makes chemotaxis sound like a simple compass—stronger attractant $\rightarrow$ move that way—but *Physarum* rarely encounters a clean, steady slope of chemical concentration. In nature, gradients are broken by turbulence in water films, by porous soil, by competing odours, by barriers, and by time: molecules diffuse, decay, and are absorbed. Even in a laboratory dish, an attractant does not produce a perfectly smooth hill. The maze walls block the straightest route, and diffusion in a maze creates a patchwork: pockets where molecules accumulate, thin zones behind corners, little chemical eddies. The geometry of the environment becomes the geometry of hunger.

A useful way to picture this is to treat the maze not as corridors but as a network of channels through which molecules travel, slowly, by diffusion. Each turn and dead end changes how quickly attractants can reach different regions. If two oats sit at opposite ends, each generates its own chemical field. Where those fields overlap, the slime mould lives in a superposition of influences. It does not need to recognize the maze as a maze. It lives inside a map drawn in concentration.

That map is not purely external. *Physarum* also changes the landscape as it moves. When it crawls, it leaves behind a thin film of extracellular slime. In the early 2010s, experiments by Chris Reid and Tanya Latty showed that the organism tends to avoid areas it has already explored, apparently using its own slime trail as an externalized spatial memory. This is a beguiling phrase—*memory* usually implies a brain—but the mechanism can be mundane and still profound. A trail can contain chemicals that signal, in effect, *I have been here*. If the slime mould is presented with a choice between a fresh corridor and one it has already traversed, it more often chooses the fresh one. The result is exploration without a central plan.

This matters because chemotaxis alone can trap you. Many roboticists know the classic U-shaped trap problem: if a machine follows the steepest upward gradient of an attractant, it can enter a U-shaped cul-de-sac and become stuck near the attractant source, unable to find the route around the barrier because every small step away looks like moving downhill. A slime mould that only climbed gradients could suffer similar frustrations in complex environments. A tendency to avoid one's own trails makes it easier to back out and try alternatives. The environment becomes annotated by the organism's past movement, like chalk marks on a labyrinth wall.

Chemotaxis, then, is not a single algorithm. It is a family of coupled processes: receptor binding, intracellular signaling, rhythmic contraction, cytoplasmic streaming, growth and retraction, and environmental modification. The remarkable part is that each of these processes can remain local. No molecule needs to carry a complete map of the dish. No region needs to know why the whole organism is changing shape. Yet when these mechanisms run together, the organism's movement begins to look goal-directed.

A historical anecdote helps here, because *Physarum* has been offering hints about distributed problem-solving for longer than its recent fame suggests. In the mid-twentieth century, before the slime mould became a celebrity of maze videos and network diagrams, it already served as a laboratory workhorse for cell biologists studying streaming and contraction. Researchers would cut plasmodia, graft them, place food at different points, and watch how waves of contraction reorganized. The organism's behaviour could be coaxed without needing to train it; one simply arranged incentives. If you want a creature to reveal what it is, you design a world that asks a clear question.

One clear question is: how does a body explore when it has no natural boundaries except those it creates? A hungry *Physarum* placed on agar typically spreads as a thin sheet, sending out pseudopod-like fronts into multiple directions at once. This looks wasteful if you imagine a creature trying to be efficient. But exploration is expensive only if you pay attention to the final path and ignore the uncertainty at the beginning. When you do not know where food is, spreading wide is sensible. It allows the organism to sample a larger area and to catch faint attractant signals that would never be detected by staying compact.

The geometry of exploration is constrained by the body's own mechanics. If the sheet extends too thinly, it risks drying and tearing; it also risks losing coordination of flow. The organism must remain continuous: cytoplasm must still stream through its network to feed the advancing edges. This produces a characteristic compromise: wide, branching fans at the frontier, connected back to thicker veins that act as supply lines. The frontier is adventurous; the core is conservative. Even in this early stage, hunger has shape.

Now place a few oat flakes at different locations. At first, the slime mould may reach them by chance—a fan touches an oat and begins feeding. Once feeding begins, chemotaxis sharpens: the local attractant and nutrient uptake reinforce flow to that region, and the fan thickens into a vein. The organism does not merely sit on the food; it reorganizes around it, building a transport network that can shuttle resources from one part of its body to another. If multiple foods exist, the organism faces a constraint that is almost mathematical: it must connect these sites while managing the cost of maintaining tube length.

That cost is real. A tube is living tissue: it must be kept pressurized, repaired, and supplied. A long tube that carries little flow is not merely unnecessary; it is a liability. The simplest way to think about *Physarum* is to picture it as an adaptive plumbing system. Flow makes pipes worth keeping. Lack of flow makes them wither.

Chemotaxis provides the bias that begins this adaptation, because attractants set up the initial asymmetry in flow. But once tubes exist, the organism enters a different regime. Within its own body, it can create internal gradients of pressure. Cytoplasmic streaming is not random sloshing; it is coordinated by peristaltic waves that travel along the network. Where a wave pattern produces higher net flux, those tubes deliver more nutrients and signals; where net flux is low, the tube starves and thins. In a maze, as the organism expands into multiple corridors, each corridor becomes a candidate conduit. The ones that offer a better connection between food sites will tend to carry more flow, for reasons that are partly geometric and partly chemical. Chemotaxis is the spark; hydraulics does much of the selection.

Still, chemotaxis remains the anchor. Without chemical gradients, the organism's exploration would have no reason to stop at the right places. The maze would be a neutral scaffold, and the slime mould would merely spread until it hit the dish's limits. To make a maze into a question, experimenters place food at particular points. That act creates two chemical lighthouses. Between them, the organism experiences a strange tension: any branch that helps connect the lighthouses becomes favoured; any branch that does not becomes dead weight.

To appreciate how local the process can be, consider a corridor junction. One branch turns toward a region of slightly higher attractant concentration; the other turns toward slightly lower concentration, perhaps because a wall blocks diffusion or because the path is longer. At the junction, both branches initially receive some cytoplasmic flow. But the branch with the stronger attractant signal at its leading edge will tend to grow more, creating a slightly wider path for flow. A slightly wider tube carries more flow; that increased flow delivers more resources and can reinforce further widening. A small bias becomes a large difference. A junction becomes a decision point without anyone deciding.

The phrase *biased random walk* is sometimes used for chemotactic movement in bacteria, where cells tumble and run, and chemical gradients modulate how often they tumble. *Physarum* is different in form but similar in spirit. Its exploration is messy, branching, and opportunistic; chemotaxis biases which branches prosper. The result can look almost intentional, like a hand tracing a route, but the underlying process is closer to a population of competing protrusions connected by shared circulation.

A concrete example makes this feel less abstract. Picture a petri dish with a moist agar base. A small piece of *Physarum* is placed near the center. Two oat flakes are placed far apart. At first, the organism spreads in all directions, a pale yellow fan. After an hour or two, one part of the fan bumps into an oat. That local region thickens, becoming brighter and more rope-like. The rest continues exploring, but the vein to the first oat becomes a stable highway. Later, another branch reaches the second oat. Now the organism has two strong sinks for nutrients and attractants. Veins that help transfer resources between these sites become important; veins that wander off to nowhere become expendable. Over time, a thick connection between the oats emerges, while the peripheral fans recede, leaving translucent slime traces like the footprints of an abandoned city.

This is hunger behaving like a geometer. It measures by sending out feelers, but the measure is not a number. It is thickness, flow, and persistence. A path that is short, unobstructed, and chemically rewarding becomes physically embodied as a prominent vein. A path that is long, blocked, or chemically poor fades.

The beauty of the maze experiments is that they force this geometry to be visible. A maze constrains diffusion into corridors. It also constrains the organism's body into discrete channels, turning a living sheet into a set of candidate edges. In the open dish example, there are infinitely many possible routes; in the maze, there are only a finite set. That makes the outcome look like a choice between options, and it tempts us to use the language of strategy. Yet the underlying driver remains the same: local response to chemical gradients, amplified and stabilized by the organism's own rhythmic transport.

It is worth pausing on what chemotaxis is *not*. It is not a GPS coordinate system. The slime mould does not know the absolute direction of the food. It does not triangulate. It does not store a map of the maze in a hidden organ. If you rotate the dish, the organism does not care; it is not oriented to north. If you redraw the maze so that corridors are different but diffusion produces similar local gradients, the organism's behaviour can look similar. Its world is not the maze as a symbolic puzzle; its world is a changing field of chemical opportunity and risk, continuously sampled by a deformable body.

In humans, hunger can be imagined, postponed, denied, or redirected. We can smell bread and decide not to eat it. *Physarum* has no such inner theatre. But it does have something that can still surprise: it can integrate information over space and time simply by being a connected, flowing material. Chemotaxis begins as a local sensitivity to molecules. The moment those local sensitivities are tied together by shared streaming, the organism becomes a kind of collective: different regions tug on the same internal circulation. A strong signal at one frontier can pull resources away from weaker fronts. The body negotiates with itself through flow.

This negotiation is why the slime mould's behaviour feels more sophisticated than that of a droplet of dye diffusing in water. Dye spreads down gradients until it evens out; it does not build enduring channels. *Physarum* is not merely responding; it is remodeling. Its hunger does not fade when it has moved a few millimetres; it persists as a structural change that biases future movement. A thickened vein is a memory of success. A slime trail is a memory of passage. In this way, the organism writes its past into its own material and into the substrate beneath it.

Seen in the right light, a maze is not a test of cleverness but a theatre where this writing becomes legible. The walls make gradients sharper, choices more discrete, and mistakes more obvious. Dead ends become places where attractant signals are weak or misleading, and where flow can stagnate. Corridors that truly connect the food sources become arteries, not because the organism has solved a puzzle in its mind, but because the chemistry of food has pulled the hydraulics of its body into a stable shape.

Hunger, for *Physarum*, is not a feeling. It is a set of couplings between outside and inside: molecules to receptors, receptors to oscillations, oscillations to streaming, streaming to growth. The geometry is drawn by diffusion and constrained by walls, but it is traced by a living, pulsing material that cannot help but follow the slopes it inhabits. Where the chemical air is richest, the organism thickens, and where it thins, it withdraws, leaving behind the faint sheen of a route once tried and then refused.

3.2 Conquering the Labyrinth

A maze, in a laboratory, is a provocation. It takes an everyday desire—food—and wraps it in geometry, forcing any would-be forager to pay a price for mistakes. For *Physarum polycephalum*, the provocation is especially stark because the organism cannot simply pick itself up and leave. It must remain continuous, spreading and retracting like a living film. When Toshiyuki Nakagaki and his colleagues Hiroyasu Yamada and Ágota Tóth carried out their now-famous maze experiment around 2000, the drama lay not in a sudden flash of insight but in a slow, almost theatrical change of form: from an indiscriminate occupation of every corridor to a single thick line between two food sources.

The set-up is easy to picture, which is part of why it has become iconic. A plastic maze, the kind you might find in a child's puzzle book, is pressed into agar so that the corridors become shallow trenches of moist, nutrient-poor gel and the walls become barriers. Two small piles of food—often oat flakes—sit at opposite ends. *Physarum*, initially placed so that it can enter the maze, is given time and darkness. Then the waiting begins.

At first, the organism behaves like a spill. It does not probe politely; it pours itself forward. Yellow fans press into every available corridor, splitting at junctions, flooding dead ends, and making the maze look less like a puzzle than like a canal system filled by a rising tide. To a human observer, this stage is oddly unsatisfying. If one expects "maze solving" to mean choosing the correct path early, then the slime mould seems to be cheating by trying everything at once. Yet this is precisely the point: without a nervous system to simulate alternatives, the organism embodies alternatives physically. The maze becomes filled with parallel hypotheses.

The transformation begins only after this saturation. Once both oat piles are contacted and feeding is underway at each end, the organism's priorities change. Parts of the maze that were initially explored begin to thin. The bright, vigorous yellow fades in cul-de-sacs. Branches that once looked as thick as the main corridors become translucent threads, then break. Over hours, the maze empties itself of slime mould except for a reinforced connection between the two food sites. The final pattern, in the simplest mazes, often coincides with the shortest route permitted by the walls.

This slow withdrawal is where the experiment earns its reputation, because it is visually unmistakable. A dead end that was once occupied is later abandoned, as if the organism had learned that it leads nowhere. A long looping route that initially seemed as plausible as any other loses its substance, while a more direct corridor thickens into a dominant vein. The effect is not subtle, and it has an almost moral clarity: wasteful extravagance gives way to thrift.

One reason the result captured attention beyond biology is that it resembles what human engineers do when asked to connect two points efficiently. If you were laying a cable through a building, you would not thread it through every corridor and then remove the extra. You would plan. Yet *Physarum* does something like planning without ever stepping outside the material act of growing. The maze is not solved by representation; it is solved by transformation.

The experiment also has a small historical irony. The late twentieth century was an era of intense fascination with computation: not only silicon chips, but also neural networks, genetic algorithms, and all manner of "natural" computing metaphors. In that atmosphere, an amoeboid organism that could trace a shortest path through a maze seemed to offer a living parable. But Nakagaki's demonstration was not a stunt performed with exotic equipment. It depended on careful husbandry—keeping the agar moist, the temperature stable, and the organism healthy—and on choosing a geometry that would make the outcome legible. The maze is a stage set designed for human eyes, and the organism performs best when its own sensory world is respected: darkness, humidity, and food that diffuses attractants.

To keep the story honest, it helps to be precise about what is meant by "shortest path." In a maze with uniform corridor widths and no special gradients except those created by the oats, the physical length of a corridor segment is a good proxy for its "cost." A shorter route generally requires less tube to maintain and can deliver nutrients faster. Under those conditions, the final thick vein often aligns with the minimal-length connection, because any longer alternative is penalized by both maintenance cost and slower transport. But the maze is not merely geometry; it is also chemistry. If one corridor is slightly more illuminated, slightly drier, or subtly contaminated, the organism may prefer a longer but safer route. The shortest path, for *Physarum*, is not a theorem; it is a compromise written in living tissue.

A concrete way to see this compromise is to imagine two routes between the oats: Route A is shorter but includes a narrow choke point where the organism has difficulty maintaining a robust tube; Route B is longer but allows a thicker, steadier connection. In purely geometric terms, A wins. In the organism's terms, B may carry higher net flux because its tube can widen. In such cases the final pattern can select B, not because the slime mould is bad at geometry, but because it is optimizing something closer to "effective transport" than "Euclidean length." This is the crucial point for interpreting the maze experiment: the organism is not aiming at a human definition of optimality. It is trying to keep food moving through its body in the easiest possible way.

Nakagaki's experimental choreography makes this tension visible by starting with a dramatic over-commitment: occupying the whole maze. That initial state is not what one would find in nature, where a slime mould would rarely have the opportunity to fill every corridor before touching food. It is, however, an ingenious way to separate two processes that otherwise blur together: exploration and pruning. By ensuring the organism has access to many routes at once, the experiment reveals how selection can occur *after* exploration, by eliminating options rather than choosing one at the start.

Watching the time course is like watching a city's traffic patterns harden into infrastructure. In the beginning, every street is congested: cytoplasm streams through many corridors, and the organism's rhythmic contractions pulse across a sprawling sheet. But the flows are not equal. Even early on, some corridors carry more cytoplasm than others, and that difference can be amplified. Where flow is steady, tubes thicken; where flow is sporadic, tubes thin. The organism does not need to know that a corridor is a dead end; it only needs to register that, over time, a dead end does not contribute to sustained exchange between the food sources. The lack of sustained exchange shows up as low net flux, and low net flux is paid for in atrophy.

This is also why the experiment is compelling to anyone who has ever tried to find their way in a confusing building. Human maze-solving often involves the same two-phase pattern: a burst of exploration, then a pruning of the remembered map as useless routes are discarded. The difference is that humans prune in memory, while *Physarum* prunes in flesh. The memory is externalized as tube thickness.

A particularly vivid moment in the maze time-lapse occurs when the organism begins to "let go" of a corridor. The tube does not simply vanish; it becomes stringy, then develops weak points where streaming stalls. Eventually it snaps, leaving behind a faint smear. The living material that once filled the dead end is reabsorbed and redistributed, contributing to the reinforcement of routes that remain. Retreat is not merely abandonment; it is recycling. The slime mould reclaims its investment.

The following diagram captures the broad arc of this transformation. The geometry is schematic rather than a faithful reproduction of a particular commercial maze, because the key is the change over time: pervasive occupation, selective retreat, and final consolidation.

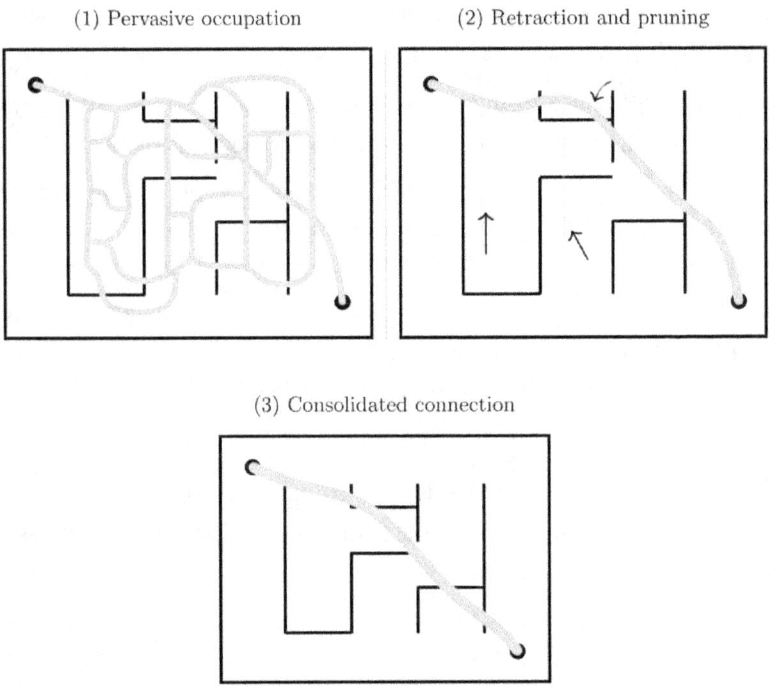

Yellow indicates *Physarum* tubes; black lines indicate maze walls.
Arrows mark retreat from cul-de-sacs.

The diagram is a simplification, but it is faithful to the most important aspect of the real time-lapse: the organism does not "find" the correct route and then follow it. It occupies broadly and then sculpts. This matters because it changes how one should interpret the behaviour. The intelligence, if that is the word, is not in choosing a direction at a junction. It is in the capacity to maintain many tentative commitments simultaneously and then let the environment and the organism's own transport dynamics adjudicate among them.

One can even feel, in the patience required to run the experiment, the distance between human and slime mould time. For a person staring at a petri dish, hours can feel like an eternity. For *Physarum*, hours are the tempo of its body: the period over which it can redistribute biomass, reinforce a tube, and erase a mistake. This is why the most striking photographs of the maze experiment often compress time into a few frames. The human eye wants a before-and-after. The organism lives in the in-between.

There is also a quiet constraint that experimenters learn quickly: the maze has to be built in a way that the organism can physically inhabit. Corridors must be wide enough to allow a tube, moist enough to sustain it, and not so deep that oxygen becomes limiting. If the agar dries at the edges, the organism may avoid the perimeter regardless of geometry. If the oats are too large, they can flood the maze with attractant so strongly that growth becomes more directly guided by the gradient, reducing the dramatic "fill and prune" sequence. The behaviour is robust, but it is not magical; it depends on the physics of diffusion and on the physiology of the organism.

This dependence helps explain why the maze demonstration, though famous, is best thought of as one carefully tuned window into a broader set of behaviours. Under some conditions, *Physarum* can be strongly guided by chemoattractant fields and may appear to solve a maze in a more direct fashion, advancing preferentially along corridors that carry stronger attractant concentrations. Under other conditions, especially when gradients are weaker or more ambiguous, it spreads more uniformly and then prunes. In both regimes, the same basic motive force operates: local growth biased by chemical cues, coupled to the organism's ability to strengthen some routes and abandon others.

The historical significance of Nakagaki's experiment lies in what it made undeniable. Before this, it was already known that *Physarum* could build networks and respond to stimuli, but these facts could be treated as curiosities of cell motility. A maze is culturally loaded: it is a symbol of reasoning. Placing a brainless organism in a maze forces an uncomfortable question. If the final pattern matches a solution that, in a textbook, would be produced by a shortest-path algorithm, where is the algorithm implemented? In a creature like a rat, one looks to neurons. In *Physarum*, the only available substrate is the body itself.

The discomfort can be productive if it is handled carefully. One can grant the organism its achievement without attributing to it a ghostly inner navigator. The maze experiment is not a proof that slime moulds think like humans. It is evidence that certain kinds of problem-solving can be performed by systems that are distributed, embodied, and relentlessly physical. The "computation" is smeared across the maze in strands of protoplasm, and the intermediate states matter as much as the final line.

A final detail often missed in popular retellings is that the organism's solution is not merely a drawn line. It is a working piece of infrastructure. If you could shrink yourself down and stand inside the thickened vein, you would not see a static trace; you would feel a steady surge of streaming cytoplasm, driven by contractions that can propagate as waves. The vein is not a record of where the organism went. It is the path along which the organism is actively moving its resources, keeping both food sites within one integrated body. The maze has been conquered not by escaping it, but by turning it into a circulatory system that makes detours unaffordable and dead ends irrelevant.

3.3 The Hydraulic Computer

The final thick vein in the maze looks like a line drawn with intent, but it is better understood as a piece of plumbing that has survived a round of ruthless budgeting. *Physarum polycephalum* has no neurons to vote on which corridor matters. Instead, it has tubes, pressure, and a ceaseless back-and-forth streaming of cytoplasm. The "decision" is a redistribution of flow, and the "memory" is a change in diameter.

A slime mould plasmodium is often described as a single giant cell, which is true in the literal sense: it is one continuous cytoplasm enclosed by membrane, containing many nuclei. But the phrase can mislead, because it encourages the picture of a uniform blob. In practice, the plasmodium builds a network. Thick, rope-like tubes branch into thinner tubes, which spread at the margins into a sheet. Inside the tubes, cytoplasm streams in alternating directions, driven by waves of contraction in the tube walls. In the sheet, the material creeps and fans, sampling the environment. The network is not an accessory; it is the organism's way of being large without becoming incoherent.

What makes the network computational is not that it resembles a circuit diagram, but that it obeys a set of physical constraints that *force* trade-offs. Cytoplasm behaves like a fluid. Tubes behave like pipes. If you drive fluid through pipes, the distribution of flow depends on pressure differences and resistance. Resistance, in turn, depends strongly on the tube's radius. This is the insight that turns slime mould behaviour from a charming curiosity into a serious model of distributed problem-solving.

For a thin cylindrical pipe with smooth walls and steady laminar flow, classical fluid mechanics gives the Hagen–Poiseuille relationship: the volumetric flow rate Q is proportional to the pressure drop ΔP

and to the fourth power of the radius r,

$$Q \propto \Delta P r^4.$$

The details involve viscosity and length, but the most important feature is the r^4. Double the radius, and you do not merely double the flow; you increase it sixteen-fold. Even modest thickening can dramatically lower hydraulic resistance. In a living network, where thickness can change, this creates the possibility of strong positive feedback: a tube that carries slightly more flow can become slightly wider, which allows it to carry much more flow, which encourages it to widen further.

This is the hydraulic core of the maze story. The maze does not need to be “understood”; it needs to be *traversed by flow*. When the slime mould initially occupies many corridors, the network contains multiple routes between the food sources. Cytoplasm can, in principle, stream along many of these routes. But the distribution will not be equal, because some routes are longer, narrower, or more tortuous, and because the organism’s own contractions impose pressure patterns that interact with the network’s geometry. Flows tend to concentrate where resistance is lower. Those same flows deliver nutrients and signals that sustain the tube walls. Over time, high-flow tubes are reinforced and low-flow tubes are allowed to decay. The computation is an iterative physical process: redistribute flow, adjust diameters, redistribute again.

Atsushi Tero and collaborators captured this logic in a simple but powerful mathematical model in the late 2000s. They treated the slime mould’s network as a graph: junctions are nodes, tubes are edges. Each edge has a conductance (the ease with which flow passes), and flow follows something like Ohm’s law for fluids: pressure differences drive flux through edges in proportion to conductance. Then comes the biological twist: conductance is not fixed. It adapts. Edges that carry more flux increase their conductance; edges that carry little flux decrease. With only this rule, the model can reproduce the pruning seen in maze experiments, leaving a dominant connection between food sources. The slime mould, in this view, is not solving a maze by searching symbolically; it is implementing an adaptive transport network.

The appeal of this picture is not only that it fits the photographs. It also explains why the behaviour is robust. Many details of *Physarum* physiology are complicated and context-dependent, but the feedback between flow and tube thickness is a generic physical mechanism. Any system that can vary its channel widths based on how much material passes through them will tend to develop a backbone of efficient routes. Engineers see this same principle in river deltas, where channels that carry more water erode and widen, capturing more flow. The difference is that a river does not care about reaching food; *Physarum* does, and it can actively resorb tissue in places where it is no longer useful.

The word *computer* can feel like a provocation here, because it conjures silicon chips and deliberate design. Yet it is not wrong if it is used carefully. A computer is a physical system that maps inputs to outputs in a reliable way. In the maze set-up, the inputs are the placement of food sources, the geometry of corridors, and the organism's physiological parameters. The output is a stabilized network pattern that connects food sources efficiently. The mapping is not instantaneous; it unfolds over hours as the organism performs the computation through growth and retraction. The "program" is not a sequence of instructions but a set of coupled feedbacks embedded in tissue.

There is an intuitive way to see the computation without writing down any equations. Imagine walking through the maze with a roll of hose, and you are told that two points must be connected so that water can circulate between them. At first you might unspool hose into every corridor to be safe. Then you turn on the pump. Water will flow more strongly along some routes than others, especially those that offer less resistance. Now suppose the hose has a special property: sections carrying strong flow automatically thicken and become smoother inside, while sections carrying weak flow become kinked and narrow. After a while, only the best route remains functional, because the others have self-sabotaged. That is essentially the logic of the slime mould, except that the hose is alive, and thickening and thinning are cellular processes.

A question immediately follows: where does the pressure difference come from? A passive network of pipes does not pump itself. *Physarum* pumps via peristalsis, a traveling wave of contraction along the tubes. Under the microscope, the tube walls periodically contract and relax, squeezing the cytoplasm and driving it forward, then reversing. The streaming is often described as "shuttle" flow because it oscillates back and forth, rather than moving in one constant direction. But oscillatory flow can still transport material effectively over time, especially when combined with asymmetries in the network or in contraction patterns. It can also serve as a signaling system: chemicals carried by the flow are distributed through the organism, allowing distant regions to influence one another without any dedicated nerves.

Karen Alim and colleagues (2013) showed that these peristaltic waves are not arbitrary. The contraction pattern organizes flows across the entire individual, and the wavelength of the dominant contraction wave can scale with the size of the organism. This scaling matters because the network is a distributed body: to coordinate far-flung regions, the pumping must be coherent at the right scale. If the wave is too short, different regions can become hydraulically out of sync; if it is too long, the pumping becomes sluggish. The organism appears to tune its contractions in ways that maximize transport.

This tuning makes the maze behaviour more plausible. A maze-filling plasmodium is not a set of isolated branches making independent choices. It is a single hydraulic entity. When one corridor begins to carry more flow, it can affect pressure patterns elsewhere. When a dead end is pruned, the hydraulic load on remaining routes changes. The computation is not local in the sense of being confined to one junction; it is distributed through the coupling of pressure and resistance across the whole network.

One can glimpse the power of this coupling in a simple mental experiment. Consider two parallel routes connecting the same endpoints. One is slightly shorter. In a fixed network, the shorter route has lower resistance, so it carries slightly higher flow. Now allow diameter to adapt based on flow. The shorter route thickens, reducing its resistance further, increasing its flow, thickening more. The longer route experiences the opposite: its share of the flow drops, so it thins, increasing resistance, losing more flow. The system amplifies an initial geometric advantage into a decisive anatomical difference. This is why a maze can be "solved" without any explicit comparison of lengths. The network does not measure length; it feels resistance through flow.

At this point, it is tempting to over-romanticize the outcome as perfect optimization, but the organism's solutions are better described as workable compromises shaped by physics and ecology. Tero and colleagues later explored how the same adaptive principles can balance competing objectives: minimizing total tube length (maintenance cost), maximizing transport efficiency (fast delivery), and maintaining robustness (alternative routes in case of damage). A network optimized purely for minimal length can be fragile: one cut severs the system. A network optimized for robustness may maintain loops and redundancies. *Physarum* can produce networks that sit in different parts of this trade-off space, depending on conditions. Under stress or in fluctuating environments, redundancy can be worth the cost.

A celebrated illustration of this idea compared patterns generated by *Physarum* to the railway network around Tokyo (Tero and colleagues, 2010). The point was not that the slime mould "designed" the railway, but that both systems face similar constraints: connect many sites, move resources efficiently, and remain functional under failures, all while paying for infrastructure. Human engineers negotiate these trade-offs through planning meetings, budgets, and political arguments. *Physarum* negotiates them through flow reinforcement and decay. It is hard not to feel a certain humility when the same tensions show up in a piece of living yellow tissue on agar.

A concrete example brings the hydraulic logic back down from metaphors and comparisons. Place three food sources at the corners of a triangle. A naive expectation is that the organism will connect all three with the shortest total length, a kind of minimal spanning tree. In practice, *Physarum* often forms a network that is close to that ideal but may include an extra loop. The loop looks like waste until you imagine an injury: if one tube is severed, the loop provides an alternate route for flow. From the organism's standpoint, this can be insurance. In a maze, the pruning can appear to remove all redundancy because the experimental goal is to highlight a single path, but in more open environments the organism's networks can preserve alternative routes when conditions reward resilience.

The hydraulic computer also explains why seemingly minor experimental details matter. A slightly drier corridor increases viscosity effects and makes streaming less effective, raising resistance and discouraging reinforcement. A narrow corridor limits the maximum tube radius, capping the potential r^4 advantage. A stronger attractant gradient can bias initial growth so strongly that certain routes receive early reinforcement even if they are not geometrically optimal. These are not "confounds" in the sense of ruining the story; they are part of what the story is really about. The organism's behaviour is the outcome of coupled fields: chemical gradients outside, pressure and flow inside, and tissue adaptation bridging the two.

There is also a subtle conceptual payoff. Many discussions of intelligence assume that information must be represented in symbols somewhere—in neurons firing, in bits stored in memory, in internal maps. *Physarum* demonstrates a different possibility: information can be stored as *impedance*. A thick tube is not merely a consequence of past flow; it is a persistent change that shapes future flow. The network itself becomes a record of what has worked. The organism "knows" the good route in the same way a well-worn footpath "knows" where people have walked: not by awareness, but by altered structure.

This idea can be made almost tactile. If you run your hand over a wooden staircase in an old building, you can sometimes feel a shallow groove where countless feet have stepped, compressing the wood slightly more in one place than another. That groove influences later footsteps, guiding them into the same channel. No one planned the groove; it emerged from repeated use. In *Physarum*, the groove is a vein. The staircase is alive and constantly repairing itself, deepening some grooves and sanding away others.

It is also worth recalling that the organism's internal streaming is not a gentle drift. Under magnification, it looks vigorous, as if the entire network is breathing. This is not the quiet accumulation of small chemical changes; it is an active, energy-consuming process. The plasmodium spends metabolic energy to pump its cytoplasm, and it spends energy to remodel its tubes. The optimization is not free. That cost is part of why pruning happens at all. If maintaining an unused branch were cheap, there would be little reason to abandon it.

The elegance of the maze outcome, then, is the visible endpoint of a continuous negotiation between expenditure and return. A tube that does not pay for itself in delivered resources becomes thinner, then disappears, its material returned to the common pool. A tube that pays for itself becomes an artery. The maze walls merely make the negotiation legible by limiting the set of possible arteries.

One can watch this negotiation not only in the final pattern but also in the organism's rhythm. Near food sources, contraction frequencies can change; along dominant veins, streaming can become more coherent. Even without understanding all the biochemical details, it is hard to miss the sense that the organism is coordinating itself through its own hydraulics, distributing not only nutrients but also signals that determine where growth should be supported.

In the end, the most unsettling lesson of the hydraulic computer is how little is required for something that looks like calculation. A pressure field plus adaptable resistance is enough to select efficient paths. There is no need for a central controller, and no need for a miniature maze in the organism's head. The maze is computed by being inhabited: the corridors compete as conduits, and the winners are the ones that can keep the cytoplasm moving with the least struggle, until the organism's body itself becomes the solution.

Chapter 4

The Urban Planner

Can a brainless organism out-design civil engineers? By placing food sources on a map of Tokyo, researchers watched as Physarum grew a network strikingly similar to the actual railway system, yet optimized for a different kind of survival. This narrative explores the tension between construction cost, transport speed, and the critical ability to withstand disaster.

4.1 Mapping the Metropolis

Tokyo is the sort of city that makes planners reach for metaphors. From above, the Greater Tokyo area looks less like a single settlement than like weather: density thickening, thinning, and thickening again as the land folds around rivers and hills, as coastlines become reclaimed ground, as suburbs stitch themselves to the centre. Movement is what holds it together. Rail lines, in particular, act like the city's circulatory system: not one grand avenue, but many branching corridors, threading neighbourhoods to hubs, hubs to hubs, and ultimately to the dense core.

That is the image Atsushi Tero and colleagues had the audacity to shrink to the size of a dinner plate. Their wager was simple and unsettling: if a city's transport network is, at heart, a solution to a recurring problem—how to connect many important places cheaply, quickly, and safely—then perhaps a creature that has been solving its own transport problem for hundreds of millions of years might, under the right conditions, stumble into a similar shape.

The creature was *Physarum polycephalum*, the true slime mould that, in its feeding stage, becomes a single cell spread out into a many-branched sheet called a plasmodium. It has no brain, no nerves, no central command. Yet it builds a network of tubes through which its cytoplasm flows. Those tubes are not decorative; they are infrastructure. Food, chemical signals, and the very pressures that drive internal circulation are routed through a living web that is constantly rebuilt as conditions change.

The experiment's genius lies in how little it asks of the organism, and how precisely it translates a metropolitan map into the mould's own language. A rail planner thinks in stations and terrain. *Physarum* thinks in food and danger.

Stations as oat flakes. The "stations" were oat flakes placed on an agar gel. Oats are more than bait: they are a concentrated, reliable food source, a point that the organism will not merely visit but attempt to keep supplied. Each flake was placed at a location corresponding, in a scaled-down way, to a major urban centre in the Greater Tokyo region. If a flake represents a city, then a tube between flakes represents a maintained route for delivering resources.

Terrain as light. On a real map, water, mountains, and other obstacles matter. In the petri dish, Tero's group used illumination as a proxy for geographical constraint. *Physarum* avoids light. Not as a philosophical stance, but because light is dangerous: in bright conditions the organism risks drying out and being damaged. A lit region becomes, to the mould, a hostile landscape it will cross reluctantly or not at all. By projecting light patterns onto the dish, one can carve out "lakes" and "mountains" without raising a single grain of sand. Darkness becomes the buildable land.

What results is a kind of translation between worlds. The map is not smuggled in by clever geometry alone. The organism is given reasons—its reasons—to treat some places as valuable and some routes as costly.

To make that translation work, the protocol needs to be careful about time, because *Physarum* behaves differently at different moments. Anyone who has watched time-lapse footage of the plasmodium has seen the rhythm: first a broad, almost reckless exploration, then a tightening. It fans outward like spilled paint, then gathers itself into thicker cords, abandoning side streets and strengthening main roads. If you freeze the process too early, you get a messy sprawl. Too late, and you may miss intermediate alternatives that were tried and discarded.

The researchers began by placing the slime mould at a designated starting point corresponding to the Tokyo area's core. From there it had a choice: to spread into the dark regions, toward the oat flakes, while skirting the lit hazards. As it crawled, it laid down a provisional mesh of thin tubes. These are the organism's speculative investments—cheap to build, easy to retract. At this stage the network is lavish and redundant, with many branches reaching toward multiple flakes and some branches going nowhere in particular. It resembles, in miniature, the early phase of surveying a route: send scouts everywhere, then decide what is worth keeping.

Then the second phase begins, and it can feel eerie to watch: the organism edits itself.

The key is flow. Inside those tubes, cytoplasm is driven back and forth by rhythmic contractions of the tube walls. Where flow becomes regular and high, the tube tends to be reinforced: it thickens, offering even less resistance, which invites still more flow. Where flow is feeble or intermittent, the tube becomes less useful and tends to shrink. This creates a feedback loop that, in engineering terms, acts like a distributed algorithm: the network reorganizes so that high-traffic routes become more efficient and low-traffic routes are abandoned.

No single part of the organism "knows" the full map. Each tube responds to what passes through it. Yet the whole network becomes, over hours, legible as a design. When many food nodes are present, the organism cannot simply choose one shortest path. It must maintain connectivity among multiple sites while still supplying each of them. It needs a compromise between thrift and reach.

Tokyo's rail network is itself a compromise of that sort, accumulated through history rather than drawn in one sitting. The first Japanese railway, opened in 1872, ran between Shimbashi in Tokyo and Yokohama, a political statement as much as a transport line: a visible pledge that the Meiji state would industrialize and connect. Over the decades, private companies and state projects layered new corridors, and the network learned, in a social and economic sense, where the flow would be. The Great Kantō earthquake of 1923 destroyed enormous swathes of the city and forced rebuilding decisions that shaped later growth; postwar reconstruction and the explosive expansion of commuter suburbs pressed the system into a dense, hub-and-spoke-and-loop arrangement. The network looks the way it does because Tokyo kept changing its mind about where "the city" truly was, and the rails had to keep up without collapsing under their own cost.

The petri dish, of course, contains none of that political drama. Yet it contains a distilled version of the same pressure: many important places, a geometry that constrains routes, and a requirement that the system keep working even as it is being built.

One of the most revealing details of the setup is what it *doesn't* assume. It does not tell the mould where to lay a "main line." It does not impose a hierarchical plan ("connect hub A to hub B first, then branch outward"). It simply marks value (oats) and risk (light) and lets the organism negotiate.

During the growth, the mould's frontier does something that resembles decision-making under uncertainty. It advances in multiple directions at once, testing passages that later prove inefficient. Some tentative routes cut across risky lit zones; they may survive if the shortcut is worth it, or vanish if the penalty is too high. Other routes arc around hazards, longer but safer, and these may persist if they form reliable corridors for distributing food-derived resources throughout the body. If one watches closely, the important thing is not the exact position of a tube at a given minute, but the pattern of *commitment*: which strands fatten into stable highways, and which fade as if they were never there.

It is tempting, when shown the final comparison, to treat the result as a parlor trick: "the slime mould recreated the Tokyo rail map." The truth is both more modest and more interesting. The organism is not copying human cartography. It is solving its own logistics problem under a set of constraints that resemble the ones humans face. Similar problems, even when tackled by radically different agents, can yield similar shapes.

A good way to feel that, rather than merely accept it, is to imagine designing a rail network under a constraint you cannot ignore: maintenance cost. Steel, concrete, labour, land acquisition, signalling, rolling stock, all add up. Minimizing cost pushes you toward sparse networks: as few kilometres of track as possible, as few redundant lines as you can justify. But pure thrift produces fragility. A single failure can isolate an entire branch. Anyone who has been stranded by a signal fault knows the sensation: a system that looked efficient on paper becomes, in practice, brittle.

Physarum lives with a similar tension. Tube material is not free. Maintaining a thick tube means committing biomass to that structure and sustaining its rhythmic contraction. Yet a purely tree-like network, with no loops, would be risky: sever one tube and a food source is cut off. Even without predators, the environment is full of accidents—dry patches, damaging light, competing microbes. The organism's built form reflects a survival logic: keep paths short enough to be efficient, but keep alternatives enough to be safe.

In the Tokyo dish, that logic plays out in miniature. Many of the final tubes concentrate into a backbone linking the major oat flakes via corridors that resemble the human-built main arteries. At the same time, the mould tends to maintain certain loops or alternate connections that a purely minimal network would discard. The result is neither a tangled mesh nor a single skeletal tree.

The eeriness of the outcome depends on careful choices that are easy to miss. Which cities count as "stations" in the first place? A metropolitan region has many plausible sets of important nodes: administrative centres, economic hubs, commuter destinations, interchange points. Change the node set, and you change the "correct" answer. The same is true for rail planning: a map that only includes Shinjuku, Shibuya, Tokyo Station, and a few others would yield one kind of network; a map that includes dozens of local stops yields another. The experiment selects a particular resolution, one at which the city's skeleton is visible.

Likewise, the light mask encodes an opinion about geography. In the real Kantō region, Tokyo Bay, the surrounding mountains, rivers, and land-use patterns all matter. In the dish, these are simplified into illuminated regions that are "expensive" to traverse. This is not cheating; it is modelling. But it reminds us that even the cleanest demonstration is not simply nature versus culture. It is a constructed question posed to a biological system.

The reward for that careful construction is that we get to watch an organism perform a kind of cartographic labour. Not by reading maps, but by growing and pruning until what remains is a workable compromise. When the mould succeeds, it makes tangible a strange claim: intelligence need not arrive as a voice in the head. It can arrive as a shape in the world.

Near the end of the run, when the tubes have thickened into their stable pattern, a comparison becomes possible. The human network is a set of engineered lines drawn to satisfy commuters, geography, and economics. The biological network is a set of living tubes drawn to satisfy hunger, risk, and the physics of flow. Put them side by side and the similarities stop being poetic and start being structural: major corridors, hub-to-hub connections, a sparse backbone with selective redundancy.

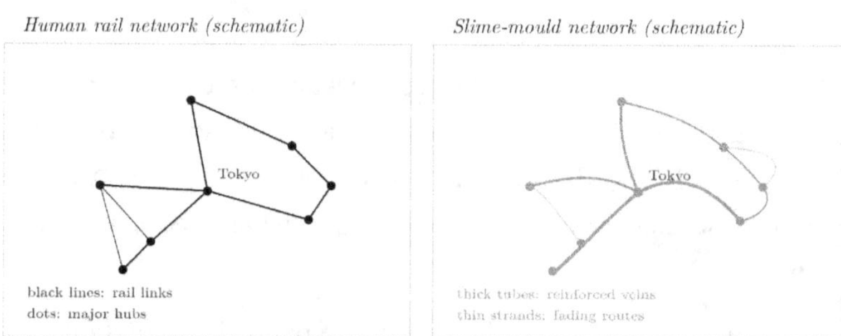

There is a particular kind of discomfort that comes from recognizing your own handiwork in another creature's solution. Rail lines feel like the signature of a modern state: budgets, land surveys, political compromise, timetables, concrete. The petri dish offers no politics, yet it does offer constraint, and it does offer a currency of cost. The slime mould pays in biomass and risk. The rail authority pays in steel and maintenance crews. Both are pushed toward designs that are cheap enough to build, short enough to be useful, and redundant enough to survive the ordinary brutality of the world.

Tokyo commuters rarely think about the network as an object that had to be *grown*. They experience it as a service: a train arrives, a gate beeps, the platform is crowded, the ride is smooth, the transfer is annoying, the timetable is merciless. The dish strips away that daily experience and leaves the skeleton behind. When a yellow living web converges on a pattern that looks familiar, it does not diminish human ingenuity. It reframes it. Some problems are so persistent—how to connect many places under limited resources, without betting everything on a single point of failure—that very different systems, built from different materials and for different purposes, can end up drawing the same lines.

4.2 The Architecture of Resilience

A transport network can be admired the way people admire a spider's web: elegant from afar, unnervingly intricate up close, and harder to repair than it looks. But the beauty is not really the point. A network is a promise that movement will continue when the world behaves badly.

The trouble is that the world *does* behave badly, in predictable ways. A storm knocks out power. A signal fails. A bridge needs emergency inspection. A construction crew finds a cracked water main. A landslide blocks a road. The question is never whether a link will fail, but which one, and when. Designing for resilience means planning for these ordinary disasters without building something so redundant and expensive that it becomes its own catastrophe.

The slime mould's tubes are made from living material, so its economics are ruthless. Every extra branch is not just length on a map; it is cytoplasm committed, rhythmic contractions maintained, a surface that can dry out, a set of walls that must be kept intact. Yet in the Tokyo experiment the final pattern was not a single skeletal tree. It kept loops. It preserved alternative routes. It paid a real price for resilience.

To understand why, it helps to borrow a small piece of graph theory. A *graph* is simply a set of points connected by lines. The points (called *nodes* or *vertices*) might represent stations; the lines (called *edges*) might represent rail segments or, in the mould, tubes. Graph theory gives names to different kinds of "good" networks, and those names become a way of talking about trade-offs without relying on intuition alone.

Begin with a harsh ideal: connect all stations using the least total length of track possible. That ideal has a precise mathematical form. If each edge has a cost (often just its length), then the cheapest way to connect all nodes *without creating any loops* is called a *minimum spanning tree*. "Tree" here means what it does in everyday language: branches, but no cycles. There is exactly one route between any two nodes. If you trace a path and ever return to where you started, you have made a loop, and a tree forbids that.

The appeal of a minimum spanning tree is obvious. It is thrift made geometric. If you had to connect a handful of research stations in Antarctica with as little cable as possible, and you could tolerate a single route between any two, a tree would feel like common sense.

Its flaw is also obvious once you picture the word "tree" literally. Chop the trunk and everything above it dies. In a minimum spanning tree, remove one edge and the graph splits into two disconnected parts. A "single point of failure" is not an accident; it is guaranteed. You can even say this as a dinner-party theorem: in a tree, every edge is a bridge.

A city does not accept this bargain. Even a small commuter line cannot be designed so that one broken segment strands an entire region with no alternative route. At the scale of Tokyo, the risk becomes political as well as logistical. A train network is not merely steel and scheduling; it is the promise that millions can reach work, school, hospitals, and one another. When that promise fails repeatedly, the city's social metabolism falters.

So why not build the opposite extreme: connect everything to everything, a dense mesh where every station has many redundant links? Because the cost explodes. Every added edge is maintenance, land, materials, operations. Even in biology, redundancy has to be selective; a body does not grow an extra aorta "just in case." It builds the right number of backup pathways in the right places.

The interesting territory lies between these extremes: networks that are almost trees but not quite, networks with *cycles* (loops) placed where they matter most. A cycle is what gives resilience its simplest mechanical meaning. In a looped network, cut one edge and flow can detour the other way around the loop. The network does not need to be fully connected in every direction; it needs to avoid being *fragile*.

The slime mould's behaviour makes this compromise feel less like a compromise and more like a habit. The mould does not "decide" to add redundancy in a strategic meeting. It arrives at loops as a by-product of how it learns where flow is. When it first explores, it creates an embarrassingly redundant mesh. That exploratory phase generates cycles everywhere. The pruning phase removes many of them, but not all. Some loops remain because, in the organism's internal traffic, they are still useful: two food sources both pull resources, and two corridors both carry substantial flow, so both corridors are reinforced rather than one being eliminated.

A minimal tree cannot express this kind of mutual usefulness, because a tree cannot contain two distinct routes between the same regions. The mould's living algorithm can, because it is not optimizing a single number like "total length." It is responding to a set of pressures that pull in different directions: shorten routes, keep transport fast, avoid risk, and keep the whole body integrated even if a tube is damaged. In engineering language, it behaves like a multi-objective optimizer. In biological language, it behaves like an animal trying not to starve.

A concrete example makes the logic tactile. Imagine five stations: one central hub and four suburban stops arranged roughly around it. The minimum spanning tree might connect each suburb directly to the hub: four spokes. Total length is low, and every suburb can reach the centre quickly. But now suppose the hub-adjacent segment of one spoke fails. That entire suburb is cut off. There is no alternative route because the tree forbids it.

Now add a single extra edge connecting two neighbouring suburbs. Suddenly, even if one spoke fails, a suburb can reach the hub by first travelling to its neighbour and then using the neighbour's spoke. One extra segment has converted a guaranteed disconnection into a detour. This is a tiny example of why loops are precious: a little redundancy can buy a lot of robustness.

What is less obvious is where to place that extra edge. Put it between the wrong pair of suburbs and it may rarely be used, becoming an expensive ornament. Put it between the right pair and it becomes a relief valve. Human planners infer "right" from ridership forecasts, land use, and a feel for where bottlenecks will develop. The slime mould infers "right" from flow itself. Where two regions exchange more traffic, the connecting tube fattens, and what fattens tends to remain.

There is a historical anecdote that shows how visceral these trade-offs can be in human infrastructure. In 1858, London endured what became known as the Great Stink. Heat and untreated sewage turned the Thames into a public-health nightmare. Parliament debated in the thick of the smell. Joseph Bazalgette's solution was not just a new sewer, but a network of intercepting sewers and pumping stations designed with spare capacity. Bazalgette chose pipes wider than strictly necessary, a costly choice at the time, and the city later benefited enormously as population grew. This was resilience in the most literal sense: a network that would not fail when the load increased beyond contemporary expectations. London did not build redundancy because it loved elegance; it built it because failure had become intolerable.

The slime mould never faces parliamentary debate, but it faces its own version of the Great Stink: local stress, drying, light, competition. A tube that carries too much flow through too narrow a conduit is vulnerable. A tube that is the sole corridor to a food source is a risk. Loops distribute load and provide alternative paths. They are, in effect, spare capacity.

Resilience also has a subtler meaning: not just surviving a broken link, but responding gracefully to threats. If one region becomes hostile, an organism does not merely need a detour; it needs the ability to *reconfigure*. *Physarum* can do this because its network is not static. Tubes can shrink, thicken, merge, and vanish. A built rail line cannot dissolve itself overnight and regrow elsewhere, but a city's transport system still undergoes slower versions of reconfiguration: service rerouting, temporary bus bridges, new lines over decades, changes in station importance as neighbourhoods evolve.

Topology—the pattern of connections—shapes how quickly such responses can happen. In experiments where the organism is forced into simple geometries, a branched form can retract from an adverse stimulus efficiently without needing to speed up everywhere. That matters because speed costs energy. If the organism can respond by local changes—strengthening contractions in a particular branch, redirecting flow through a cycle—it avoids a metabolically expensive "panic mode." A tree-like structure tends to funnel too much responsibility into too few segments; a looped structure spreads responsibility.

This is one reason loops can be seen as a form of *distributed control*. In the absence of a central brain, the organism needs a way for information to influence action across the body. Some of that influence is chemical: food sources emit attractants, hazards produce repellents. Some of it is physical: pressure differences drive streaming; contractions propagate. The network's architecture determines how these signals and flows propagate. A loop gives two pathways for influence to spread. If one path is compromised, the other remains. The organism's "decision" to abandon a region or reinforce a corridor is made by many local adjustments, but those adjustments are coordinated by global circulation.

That circulation is not a metaphor. The plasmodium's tubes undergo rhythmic contractions that drive cytoplasm back and forth, a kind of peristalsis. This is the organism's internal logistics system, and it matters because it ties structure to function. A thicker tube is not merely a line on a diagram; it is a lower-resistance conduit. Flow tends to concentrate there, delivering nutrients and signals faster. The network is both road map and traffic.

Here the minimum spanning tree becomes an even poorer model for living systems. A tree is efficient in its *materials*, but it can be inefficient in its *dynamics*. If a system's flow has to reverse, redistribute, or adapt, a tree offers few options. It channels everything through narrow chokepoints. A mesh offers multiple routes, allowing flow to reorganize without requiring complete reconstruction.

Yet the mesh is not free. Every extra link is a promise of maintenance.

In a city, this means budgets. In the mould, it means metabolically active tissue that must keep contracting and remain hydrated. The mould does not keep loops everywhere. It keeps loops where the traffic justifies the cost.

This selective redundancy shows up in careful analyses of real *Physarum* networks: the thick "backbone" segments often lie on relatively large cycles, and important segments can belong to multiple cycles. That sounds technical, but the intuition is simple. If a segment is important, it carries a lot of flow. If it carries a lot of flow, it is dangerous for it to be a dead-end dependency. Embedding it in cycles means that if part of the backbone is compromised, the organism can still route around trouble without starving a region of the body.

Human networks show the same habit when you look for it. Major ring roads and orbital rail lines are expensive, yet cities build them because they change the failure mode of the entire system. A hub-and-spoke design forces everything through the centre, making the centre fragile and congested. A ring provides an alternative way to move between spokes without passing through the hub. Tokyo's Yamanote Line, looping around the core, is not merely a convenient route; it is a piece of robustness engineering, distributing traffic and providing detours during disruptions.

The slime mould's loops are not perfect circles, and they are not planned. But they play a similar role: they prevent the network from being held hostage by any single corridor. In the Tokyo dish, one sees thickened "arteries" complemented by a few strategic cross-links that make the whole design more forgiving.

If this begins to sound like the mould is secretly a mathematician, it is worth keeping the emphasis on mechanism rather than mystique. The mould does not compute a minimum spanning tree and then decide to add redundancy. It grows a dense, exploratory network, then reinforces high-flow routes and prunes low-flow ones. In doing so, it often lands near a family of networks that engineers and mathematicians already know to be good compromises between cost and resilience. The surprise is not that such compromises exist; the surprise is that a brainless organism, using only local feedback and global circulation, repeatedly finds them.

There is also a warning embedded in the comparison. Visual resemblance can flatter us into thinking we have discovered an "optimal" design. But optimality depends on what you measure. Minimize total length and you get a tree. Minimize travel time between all pairs and you may push toward a denser mesh. Maximize fault tolerance under random failures and loops proliferate; maximize fault tolerance under targeted attacks and the story changes again. Human networks are designed under constraints that include politics, property rights, and historical accidents. The mould's constraints are different. The fact that their solutions overlap suggests a shared geometry of the problem, not an identity of goals.

Resilience, then, is not a single property. It is a balancing act carried out in the language of connections. A network that is too spare will break cleanly. A network that is too redundant will bleed resources continuously. Somewhere between the ascetic tree and the extravagant mesh lies a design that can take a hit and keep moving.

That middle ground is where the slime mould lives: not because it has a concept of robustness, but because the world punishes brittleness, and it pays, quietly and repeatedly, for just enough insurance.

4.3 Benchmarking the Blob

A network can look right and still perform badly. Anyone who has followed a beautiful-looking detour sign to a dead end learns this quickly. The eye is easily persuaded by symmetry, smooth curves, and a pleasing sense of order. Infrastructure is less forgiving. If a design is going to be taken seriously—as biology, as engineering inspiration, or simply as a claim about how intelligence can be distributed—it has to survive a colder question: compared to what, and on which scorecard?

That is where the slime mould story becomes unexpectedly modern. The moment you begin to compare a biological network with a human one, you are forced to name your values. Do you care most about building cost? About travel time? About the ability to keep moving when a link breaks? About fairness, so that outlying nodes are not treated as afterthoughts? Human transport agencies fight over these values in public, with budgets and voters. A plasmodium "fights" over them internally, with flow and starvation. Benchmarking is the act of turning those fights into numbers.

The first number is the easiest to explain, and the hardest to escape: *cost*. In spatial networks, cost is often approximated by total link length. Every centimetre of rail is expensive; every millimetre of tube is living tissue that must be maintained. Total length is not the whole story—bridges and tunnels cost more than flat track; thick tubes cost more than thin ones—but it captures a basic truth: more network means more ongoing burden.

The second number is *efficiency*, meaning how quickly or directly the network moves something from one node to another. A simple way to quantify this is by *shortest-path length*: for each pair of nodes, find the shortest route along the network, then average those distances. When that average is small, the network is, in a practical sense, well connected. People experience this as "I can get across town without endless transfers." A cell experiences it as "nutrients and signals reach distant regions without meandering."

The third number, the one that gives resilience its teeth, is *fault tolerance*. Here the idea is to ask: if a link fails, how badly does connectivity suffer? One can simulate this by removing an edge and measuring what happens to efficiency or to the number of nodes that remain mutually reachable. A robust network degrades gracefully. A brittle network snaps.

These three are already enough to generate conflict. A minimum spanning tree is spectacular on cost and terrible on fault tolerance. A dense mesh is the opposite. Real engineered networks, and many slime-mould-grown ones, live between.

Atsushi Tero and collaborators (2010) treated the Tokyo comparison in exactly this spirit: not as a single "did it match?" question, but as a trade-off surface. Their work grew out of a mathematical model Tero developed with Ryo Kobayashi (2007), which captured a key physiological feedback: edges that carry more flow become more conductive (effectively widening), while those with little flow shrink and may disappear. The model is compelling not because it imitates life down to the molecule, but because it formalizes a kind of local reinforcement that can be simulated, perturbed, and compared.

The model also offers a neat way to see why benchmarking is necessary. With a single food source and a single destination, reinforcement tends to concentrate flow onto a shortest path, and the rest dies away. That sounds like "optimality" until you remember that a living organism rarely has a single destination. With multiple food sources, reinforcement does not select one path; it negotiates among many. The network that survives is not a single line but a compromise among competing demands. That compromise can be scored.

The most honest benchmark begins with a baseline that has no biological mystique at all: geometric graphs.

Imagine placing cities as points on a plane. Without thinking about politics or biology, you can connect these points using rules that depend only on distance. The minimum spanning tree is one such rule. There are others that are slightly more generous, adding edges when points are "close enough" under various definitions. This family of proximity graphs is useful because it tells you what geometry alone can explain. If a slime mould's network resembles a particular proximity graph, then perhaps the surprise is not that it found something magical, but that geometry itself pushes most reasonable systems toward similar shapes.

These baselines become foils. If the mould does no better than a simple proximity graph on cost and efficiency, then the story becomes one of convergence rather than ingenuity. If it does better, or if it achieves a more favourable trade-off, then something about the biological dynamics matters.

Here is a concrete way to picture the trade-off without drowning in formulas. Suppose you generate many candidate networks for the same set of nodes: the minimum spanning tree, a slightly denser proximity graph, a very dense triangulation, perhaps even a few random networks. For each candidate you compute two scores: total length (cost) and average shortest-path length (efficiency). Plot cost on the horizontal axis and efficiency on the vertical axis. No single network will be best at both. Instead you often get a curve: designs that are very cheap tend to be inefficient; designs that are very efficient tend to be expensive.

The interesting designs are those that sit on the *Pareto frontier*. This is a term from multi-objective optimization. A network lies on the frontier if you cannot improve one score without worsening the other. In ordinary language: it is not dominated. You may still dislike it, but you cannot call it plainly inferior.

Fault tolerance can be added as a third axis, or treated as a second plot, or folded into the scoring by measuring how much efficiency drops under simulated damage. There are many legitimate choices, and each reveals something slightly different. That, too, is part of the point: metrics are not neutral. They express what you care about.

The slime mould's claim to fame is not that it always "wins." It is that it often lands near the frontier, in a region that human designers also seem to occupy. It finds designs that are not the cheapest possible, not the fastest possible, but hard to beat on the combined ledger.

A quick historical analogy makes the mood of this comparison clearer. In 1854, during a cholera outbreak in London, John Snow mapped cases around the Broad Street pump. The famous part of the story is the map itself, but the deeper move was methodological: he compared patterns, asked "relative to what would we expect by chance?", and used that comparison to infer a mechanism. Snow did not settle for a plausible narrative; he demanded a discriminating test. Benchmarking the blob is a similar kind of move. The question is not "is the network pretty?" but "does it outperform naive alternatives under relevant measures?"

When researchers extend the slime mould comparisons beyond Tokyo—to other national road or rail systems, to different sets of nodes, to variations in terrain constraints—the value of benchmarking grows. It prevents one dramatic demonstration from becoming a myth. A dramatic match might occur because the chosen nodes almost force a particular backbone. Or because the terrain mask acts like a strong prior. Or because human engineers, faced with the same geometry, were inevitably drawn toward similar proximity-graph-like solutions. Benchmarking against multiple baselines and multiple conditions helps disentangle these possibilities.

There is also a second kind of benchmarking that matters for real-world relevance: not the design of the network's topology (which links exist), but the distribution of traffic on a given topology. Human rail networks, for political and historical reasons, cannot be redrawn from scratch very often. They must be tuned: schedules, capacity, signalling, where to add express services, where to reinforce weak points. Physarum-inspired algorithms have been applied in this spirit too, treating reinforcement as a way to allocate flow so that heavily used routes become "stronger" while weak ones fade. In the organism, strength is tube diameter and conductivity. In an engineered analogue, it might be capacity investment.

Even if one never plans to build a railway "like a slime mould," this offers a useful mental shift. Standard engineering often begins with an explicit objective function: minimize cost subject to constraints, or maximize flow subject to a budget. The slime mould begins with dynamics: let usage change capacity, and let capacity guide usage. Objective functions appear later, when you score the outcome. This inversion can matter, because dynamic rules can produce solutions that are robust in ways an explicit optimizer might miss, especially if the optimizer is tuned too narrowly.

All of this raises a reasonable suspicion: can a simple reinforcement rule really explain the mould's performance, or does it merely imitate the outcome? The strongest argument that something real is being captured comes from physiology. The plasmodium is not an abstract graph; it is a pump-driven transport network. Rhythmic contractions generate pressure gradients; pressure gradients drive flow; flow carries nutrients and signals; those in turn influence local growth and tube maintenance. The global pattern is an emergent result of these coupled processes. The reinforcement model is a caricature, but it is a caricature that preserves the essential feedback: use strengthens, neglect weakens.

It also makes a clear prediction. If you adjust how strongly reinforcement responds to flow, you should see different network "personalities." Strong reinforcement should prune aggressively, producing sparse, tree-like designs with low cost and low redundancy. Weak reinforcement should preserve more links, producing mesh-like designs with higher cost and higher robustness. This is the kind of parameter-sensitive behaviour that engineers recognize immediately: a knob that trades efficiency for resilience is not a bug; it is a design control.

That prediction matters because it reframes the biological result. The slime mould is not a mysterious oracle that outputs the one true network. It is a system with tunable tendencies. In a lab dish, those tendencies depend on humidity, light, nutrient distribution, the size of the organism, and the history of its growth. In a model, they depend on parameters. In both cases, "the" network is not a single answer but a family of answers produced by the same underlying logic.

A Pareto plot makes this family visible. The biological network is one point. Human designs are other points. Simple baselines form a cloud. The interesting question becomes: where, in this space of trade-offs, does nature choose to live?

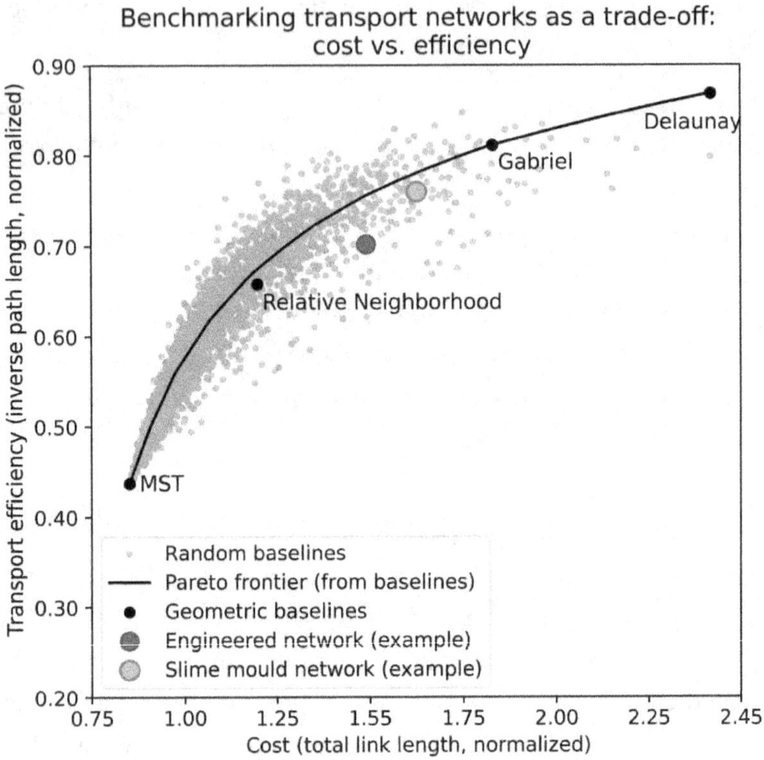

A figure like this should not be read as a victory scoreboard. It is closer to a moral map. If the slime mould sits near the frontier, it suggests its local reinforcement dynamics tend to avoid obvious waste. If it sits near the region occupied by engineered designs, it suggests that human networks, whatever their politics and history, are constrained by the same geometry of trade-offs. If, under certain conditions, the mould shifts toward sparser or denser outcomes, it suggests that resilience is not a fixed property but a choice, one that can be tuned by the strength of reinforcement and by the severity of risk.

Fault tolerance adds another twist. Two networks can have similar cost and efficiency yet behave very differently under damage. A network with a few well-placed cycles can keep most nodes connected even after random failures; a tree-like network cannot. Engineers know this intuitively when they speak of "redundancy," but benchmarking forces specificity: how much redundancy, where, and with what payoff?

This is where the slime mould's biology becomes more than a curiosity. It is not trying to impress an auditor. It is trying to keep its internal transport running in a world where damage is normal. That bias shows up in its designs. Humans often optimize for speed or cost because the environment of a rail network is, compared to a forest floor, relatively controlled. Nature optimizes for survival because the environment is not controlled, and cannot be. The mould does not treat robustness as a luxury feature to be purchased after efficiency is maximized; it treats robustness as part of what efficiency *means*.

There is a quiet human lesson in that. When infrastructure is evaluated only in good weather, it is easy to conclude that redundancy is waste. When it is evaluated across a decade of failures—storms, accidents, maintenance errors, growth beyond forecasts—redundancy becomes an investment in continuity. A living network makes that investment automatically, because the penalty for brittleness is immediate.

Benchmarking does not turn a slime mould into a superior engineer. It does something more useful: it turns intuition into comparison, and comparison into a sharper sense of what is being optimized, by whom, and at what cost. A blob on agar cannot care about commuters, but it can embody, in the thickness of its tubes and the persistence of its loops, a principle that cities relearn repeatedly: a network earns its keep not when everything works, but when something breaks and life continues to move.

Chapter 5

The Calculus of Decision

Even without a brain, Physarum polycephalum functions as a shrewd economist, constantly weighing the nutritional value of potential food sources against the perils of its environment. By observing how the slime mold negotiates illuminated barriers to reach sustenance, we discover a primal form of computation embedded entirely within the physics of fluid dynamics. This chapter reveals how simple biological oscillations enable the organism to balance speed against accuracy, mimicking the cognitive trade-offs found in higher animals.

5.1 Dining with Danger

A forager with a brain can afford to be indecisive. A forager without one has to be *decisive in the right way*, because every choice is also a commitment of body. For *Physarum polycephalum*, an acellular slime mould that spreads as a single giant cell, choosing between two dinners is not a matter of pointing left or right; it is a matter of *growing* left or right, pushing living mass down one corridor rather than another. That makes its decisions slow enough to watch and, oddly, familiar enough to recognise: the organism behaves as if it were weighing a payoff against a penalty, accepting danger when the reward is large, refusing it when the reward barely improves on the alternative.

The danger in many laboratory arenas is light. The plasmodium is nocturnal in temperament: it tends to avoid illumination, especially in the blue and near-ultraviolet range. This is not mere squeamishness. Bright, drying light can mean desiccation, oxidative stress, and in nature, exposure on the forest floor. Light is a proxy for risk. Food quality, by contrast, is usually represented by oat flakes, a kind of laboratory comfort food for *Physarum*. But not all oats are equal. Researchers can vary the concentration of nutrients infused into agar, or offer food items that differ in nutritional richness. Now the animal-like question becomes precise: when the best meal lies behind a hazard, does the slime mould take the risk?

A neat way to stage that dilemma borrows a metaphor from behavioural economics: the *two-armed bandit*. In a casino, two slot machines offer uncertain payoffs; a player must decide which arm to pull. In the slime mould's version, the arms are two paths. One path is safe but leads to poorer food. The other path is more dangerous—bathed in bright light—but leads to richer food. The mould cannot sample the reward without moving toward it, yet it can also sense trouble along the way. It is a situation built to reveal whether separate bits of information can be integrated into a single action.

One of the most compelling demonstrations came from Tanya Latty and Madeleine Beekman (2010), who set up precisely this kind of conflict between reward and risk. When both food patches were in the shade, *Physarum* behaved like a straightforward gourmet: it reliably grew toward the higher-quality food. There is nothing mysterious about that; chemotaxis (movement guided by chemical cues) can do the job. The interesting test comes when the richer option is paired with light. Under those conditions, the plasmodium does not simply flee the illuminated corridor, nor does it always lunge for the best meal. Instead, its choices shift with the *difference* in food quality. When the high-quality food is much better, the mould is more willing to brave the light; when the difference is small, it tends to take the safer route to the inferior meal. The behaviour looks like a living version of a familiar human sentence: "It's not worth it."

That sentence hides a calculation. "Worth" means some mental currency in which a reward can compensate for a cost. In *Physarum*, there is no internal accountant, but there is a body whose growth can be allocated, withdrawn, and reinforced. A corridor that is bright does not merely repel; it changes the mould's physiological state. Light can reduce advance, alter the rhythm of contraction, and in strong doses suppress the very oscillations that drive internal transport. Deciding against light is therefore not a preference in the abstract. It is an attempt to keep the organism's transport system stable enough to continue foraging at all.

A concrete picture helps. Imagine the plasmodium placed at the base of a Y-shaped track on moist agar. At the ends of the two arms sit oat flakes. The left arm is shaded and dull, leading to a meagre flake. The right arm is lit by a bright lamp, leading to a rich flake. Early on, the slime mould does what it often does: it spreads, sending exploratory fans into both corridors. Under the lamp, however, the fan hesitates, thins, and sometimes retracts. In the shade, it thickens. Yet if the rich flake is rich enough, the illuminated fan persists. Over time a tipping point is reached, not by a single command, but by the quiet accumulation of advantage in one route's physiology. The winning arm becomes a thicker tube; more cytoplasm flows through it; that flow delivers more energy and material; the tube thickens further. The decision is etched into plumbing.

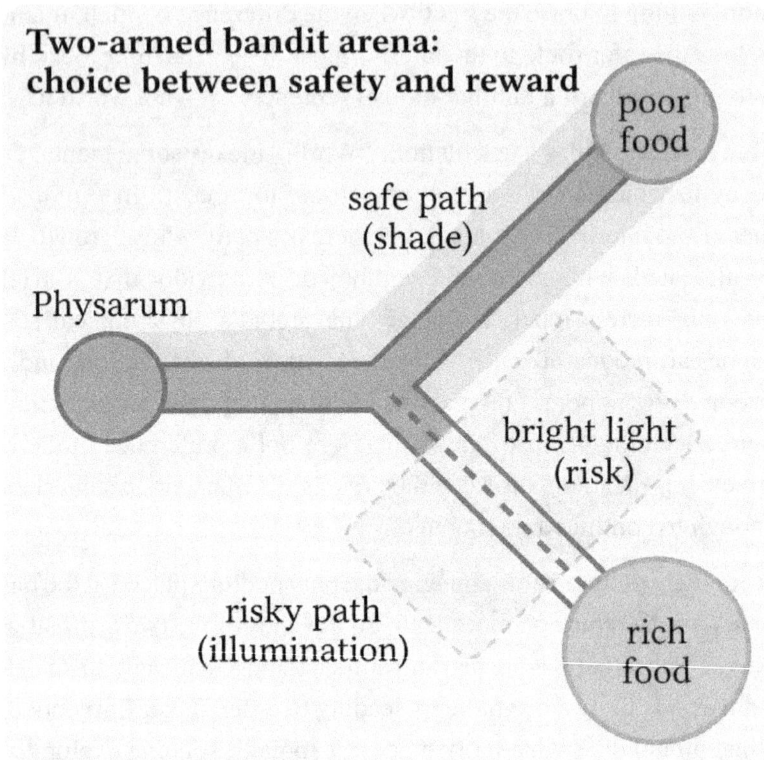

The diagram captures a crucial subtlety: *Physarum* rarely makes a clean, instantaneous choice. It samples. Those tentative tendrils are not noise; they are data-gathering, a form of embodied polling. In many animals, sampling is done with eyes and antennae while the body stays put. Here, sampling is done by the body itself. That difference forces us to take the organism's cost—benefit behaviour seriously, because the costs are paid in the currency the mould controls best: how it distributes its own mass.

A historical anecdote makes the point that such behaviour has long tempted observers to over-interpret. In the early twentieth century, as psychologists argued about whether learning required consciousness, simple organisms were repeatedly recruited as test cases. Amoebae, planaria, and even plants were marched into the lab to see whether they could "choose" or "remember". Many of those experiments were poorly controlled by modern standards, and the claims often outran the evidence. Yet the impulse was not foolish. If a system without neurons can generate flexible, context-sensitive action, then intelligence cannot be identical to brains; it must be a property that *can be implemented* in multiple materials. *Physarum* earns its place in that lineage because the decisions are not just flexible but also *graded*: the risk is tolerated in proportion to the reward.

The phrase "multi-criteria decision" can sound bureaucratic, but it names something vivid. Two criteria are being combined: the lure of nutrients and the aversion to light. Neither cue is absolute. Nutrient cues draw the mould, but only if the route is traversable and does not disrupt its internal rhythms too much. Light repels the mould, but the repulsion is not a brick wall; it is more like an added toll on the road. Whether the mould pays the toll depends on the expected gain.

To see why that matters, consider what would count as a trivial explanation. If the mould simply hated light and always avoided it, then the experiment would show only photophobia. If it always pursued the highest nutrient concentration regardless of light, then it would show only a strong chemical attraction. In Latty and Beekman's setup, the outcome depends on the *relative magnitude* of the two factors. That dependence is precisely what makes the behaviour computational in an interesting sense. It suggests that separate sensory channels are being integrated over time into a single bias on growth.

How can a single cell integrate cues without a nervous system? The answer begins at the level of physics and chemistry. The plasmodium is threaded with tubes, a network of veins that carry cytoplasm back and forth in a rhythmic process called shuttle streaming. These flows are driven by waves of contraction in the actomyosin cortex—the same basic contractile machinery that animal cells use to change shape. In *Physarum*, the contractions are organised into oscillations that travel through the network. Where the cortex contracts more strongly or more frequently, pressure gradients change; fluid is pushed and pulled; nutrients and signalling molecules are redistributed.

Now add the environment. Food does not simply attract; it *feeds* a region of the network. A tube that reaches a good food patch gains energy and materials locally. Light does not simply repel; it can stress a region, altering metabolism and contractility. Both cues therefore influence the very machinery that determines how much cytoplasm is delivered to a given branch. If a branch has better access to energy (because it reached good food), it can sustain stronger contractions and thicker tubes. If a branch is under light stress, its contractions may weaken or become less effective, reducing flow. Over time, small physiological differences are amplified into a macroscopic decision: one route becomes the main highway; the other becomes a cul-de-sac.

This is why the mould's behaviour can look rational without requiring a miniature brain. "Rational" here does not mean that it is consciously comparing options. It means that the resulting choice pattern is consistent with a cost—benefit trade-off. An economic model might describe the decision as maximising a utility function: reward minus risk. A biophysical story describes it as the coupled dynamics of growth, contraction, and flow under competing environmental constraints. The remarkable point is that both stories can be true descriptions at different levels.

There is a further twist: the cost of light is not fixed; it depends on the mould's internal state. Nakagaki and Umemura (1996) showed that photoavoidance in *Physarum* has an action spectrum—its sensitivity varies with wavelength—and that physiological conditions such as starvation and temperature can modify the response. This is what one would expect if light is not just an external command but a stressor whose impact is mediated by the organism's energetic reserves. A well-fed plasmodium can afford to be cautious, because it is not forced to gamble for calories. A starved plasmodium may accept risks it would usually refuse, because the alternative is slow depletion. Even in humans, risk tolerance changes with need; in *Physarum*, need is written into the chemistry of contraction and transport.

Put that into the two-armed bandit arena and the narrative becomes almost uncomfortably familiar. Imagine you have enough money in the bank. You take the safe job with decent pay. Imagine you are desperate. You take the hazardous work for higher pay, even if it carries costs you would otherwise avoid. The slime mould does not have money, but it has energy and time. Light can slow its spread and disturb its internal logistics; delaying too long can mean losing opportunities and drying out. The trade-off is ecological.

A second concrete example brings the point closer to nature. On a damp night in a forest, a *Physarum* plasmodium might be creeping through leaf litter that is patchy in moisture and nutrients. Under a log is darkness and humidity but limited food; across an exposed patch is a rich sprinkling of microbes but greater airflow and dryness. The mould's light avoidance in the lab stands in for a family of hazards outdoors: exposure, dehydration, and the metabolic strain of maintaining streaming in a harsher microclimate. A mould that always avoids exposure would miss valuable resources; a mould that always charges ahead would die of stress. The observed behaviour—taking risks only when rewards are sufficiently large—is not a laboratory curiosity but a plausible survival strategy.

The situation also clarifies why it is misleading to imagine the slime mould as a single decision-maker sitting at the base of the Y-junction. The organism is distributed. At the fork, it may look like a head choosing. But the real computation is carried out along the arms, in parallel. Each exploratory front experiences its local conditions: light intensity, nutrient cues, moisture gradients. Each front modifies its local contractile rhythm and tube formation. Those local changes feed back into global flow. The "opinion" of each arm is transmitted hydraulically, through pressure and streaming, rather than electrically through nerves. Consensus is a physical process: the arm that manages to sustain stronger, more coherent transport begins to monopolise the circulation, and the rest of the body is dragged, quite literally, into agreement.

That picture helps to resolve an apparent paradox in the risk experiments. How can the mould know that the risky path leads to rich food if it avoids the light? It does not know in advance; it *tests*. The tentative tendril into the illuminated corridor is like a hand hovering over a hot stove. It samples enough to detect the gradient of attractants from the rich food while also registering the stress of light. When the reward signal is strong, the tendril persists long enough to establish a tube capable of meaningful flow. When the reward signal is weak, the cost dominates and the tendril retracts. In human terms, it is not optimism or pessimism but a change in the threshold at which the organism commits resources.

There is an ethical dimension, too, though not in the sense of animal welfare. The ethical dimension is about how we talk. It is tempting to describe the mould as "brave" when it crosses the light or "cautious" when it retreats. Those metaphors can be useful, but they can also smuggle in the idea of a homunculus making choices. A better habit is to keep the metaphor tethered to mechanism: the mould behaves as if it is balancing danger and reward because danger and reward both alter the dynamics of its transport network. The choice is the emergent outcome of those dynamics.

Once that habit is in place, the slime mould becomes a teacher. It shows that cost—benefit behaviour does not require a brain; it requires a system that can (i) sense multiple features of the world, (ii) integrate those features over time, and (iii) commit resources in a way that amplifies small differences into a stable outcome. *Physarum* does all three with tubes, flows, and rhythms. The lesson is not that the mould is secretly a little animal, but that decision-making is broader than the nervous system. It is a pattern that can be built from living hydraulics.

There is a final detail, easy to miss if one thinks only in terms of behaviour. When a slime mould chooses the risky rich path, it does not simply arrive at a feast; it also remodels itself to cope with the risk. Tubes may become thicker to maintain flow under stress, or growth may proceed in a more cautious, stop-start manner. The decision is therefore not just about where to go but about what kind of body to be while going there. That is a more intimate form of calculation than most animals face. A mouse can run through a brightly lit corridor to a food pellet and remain the same mouse at the end. A slime mould that grows through the light becomes, in a literal sense, a different network: its architecture has been shaped by the very risk it tolerated.

In that light, "Dining with Danger" is not a metaphor but a description. For *Physarum*, dinner is never separate from the route taken to reach it. The meal and the hazard are folded into the same material process: protoplasm advancing, tubes thickening, oscillations adjusting, flow redistributing. The outcome can look like a preference, but it is also a record of negotiation between what the environment offers and what the organism can physically afford. The mould does not merely go toward food; it *pays* for food with the stability of its own internal circulation, and it pays only when the meal is worth the price.

5.2 Fast or Flawless?

A decision is a kind of promise to the future. For an animal with legs, a promise can be revised: step forward, step back, change course. For *Physarum polycephalum*, a decision is paid for in plumbing. To commit to one food source over another is to thicken one set of tubes, to send more of the circulating cytoplasm through one route, and to let other routes wither. That makes hesitation costly in its own way, yet haste can be ruinous. The organism faces the same dilemma that haunts everything from chess players to emergency-room doctors: do you decide quickly, or do you decide well?

Psychologists have studied this tension for more than a century, often under the name *speed—accuracy trade-off*. The phrase can sound like jargon, but the experience is universal. When you are late to catch a train, you read the departure board differently than when you are early; you act on partial information because time itself becomes expensive. In laboratories, researchers have long made this trade-off visible with reaction-time tasks: press a button as soon as you see a stimulus, or wait until you are sure. Faster responses come with more errors. Slower responses are more accurate. The fact that this pattern shows up so reliably in humans and other animals has made it feel almost inseparable from nervous systems.

Then a slime mould enters the conversation and refuses to behave as a curiosity. Latty and Beekman (2011) challenged *Physarum* with a foraging problem that is easy to describe and surprisingly hard to solve: given several food options that differ in quality, pick the best one. It sounds trivial until one remembers that the organism has no eyes to scan the scene and no brain to keep a ranked list. It must gather information by growing and sampling, while its body is simultaneously becoming the instrument of its own measurement. Under those constraints, time is not a neutral variable. Time is part of the algorithm.

Start with a concrete scene. A plasmodium sits on agar at the centre of a Petri dish. Three oat-based food sources are placed at equal distances around it, like numbers on a clock. One is high-quality, one is medium, one is poor. The mould sends out several broad fans. Each fan oozes forward, secreting enzymes, sensing chemicals, and laying down slime that marks where it has been. Eventually, one front makes good contact: it touches an oat flake, begins to feed, and a tube thickens behind it. If this were the end of the story, the mould would simply commit to whichever option it encountered first. But that would be a terrible strategy in a world where food patches vary widely in value. Sometimes the first patch is a stale crust and the better meal lies a few centimetres further.

So the organism needs a way to delay commitment long enough to compare options, while not delaying so long that it wastes energy or dries out. In Latty and Beekman's experiments, that balance shifted systematically with the mould's condition and its environment. Under stressors such as light exposure or hunger, *Physarum* tended to decide faster—in the sense that it committed to one option sooner—but its choices were more error-prone. When circumstances were gentler, it took longer and selected the best option more reliably. The pattern mirrors what cognitive scientists often find in animals: urgency compresses deliberation, and compressed deliberation raises the chance of mistakes.

It is tempting to imagine the mistake as a simple blunder, like choosing chocolate when you meant to choose vanilla. But for the slime mould the error has a bodily cost. A wrong choice means investment in the wrong tubes, diversion of streaming cytoplasm down a suboptimal route, and lost time before the organism can reconfigure itself. In ecological terms, a mistake can mean missing a fleeting patch of bacteria after a rainfall or exhausting itself before reaching shelter. Accuracy matters because the choice is literally built into the organism's architecture.

To understand why speed and accuracy are in tension, it helps to make the idea of "difficulty" concrete. In the dish, the difference between food qualities can be large or small. When the best food is much better than the second-best, even a fast-and-dirty sampling procedure will usually get the right answer: the chemical cues are stronger, the feeding payoff is larger, the local physiology changes more dramatically. But when two options are similar—two oat flakes that differ only subtly—the cues are closer together. A brief sampling can easily miss the better one, or latch onto a slightly worse option by chance. Under those conditions, longer sampling becomes valuable. Time is what allows the organism to integrate weak evidence until the best option stands out.

This sounds suspiciously like the logic used to describe decision-making in humans, where many models treat choices as an accumulation of evidence over time. In such models, a noisy stream of information is integrated until it reaches a threshold that triggers commitment. If the threshold is low, decisions are fast but error-prone. If the threshold is high, decisions are slower but more accurate. The appeal of *Physarum* is that one can translate this logic into a physically grounded story. Evidence accumulation becomes literal accumulation: more contacts with a food source, more time for nutrients to enter local tubes, more time for the contraction rhythm to adjust, more time for streaming to redistribute resources. The threshold becomes a structural tipping point: a tube becomes thick enough, or flow becomes biased enough, that one route begins to dominate the circulation.

A historical anecdote shows why this is such a powerful bridge between psychology and biology. In the nineteenth century, Franciscus Donders tried to measure the time taken by mental operations by comparing reaction times in different tasks. His work gave rise to *mental chronometry*—the idea that thinking has a measurable duration. A century later, psychologists built elaborate theories of decision time using stopwatch-like experiments. Yet it was always difficult to tie those milliseconds to underlying mechanisms. With *Physarum*, the timescales are slower, and the mechanisms are visible. One can watch the organism take time, and then see where the time went: into extra sampling fans, into repeated contacts, into thicker tubes, into a gradually shifting bias of flow.

The speed—accuracy trade-off is often described as if it were a limitation: you would like to be both fast and accurate, but biology won't let you. There is another view, kinder to the organism. A trade-off is also a *policy*. It is a way to adjust behaviour to circumstances. When hunger is severe, the cost of waiting can exceed the cost of being wrong. When conditions are safe, the cost of waiting is smaller, and accuracy becomes a better investment. The crucial point is flexibility: the organism should not have a single fixed decision time. It should have decision times that change in an orderly way with urgency.

That flexibility appears even within a species that is, in any ordinary sense, brainless. Dussutour and Ma (2019) pushed the story further by showing that *Physarum* individuals are not interchangeable. Some are consistently fast explorers; others are slower and more cautious. These differences are not just random fluctuations from day to day. They behave like stable traits, in the modest sense that an individual tends to remain a "fast" or "slow" type across repeated trials. The ecological implication is immediate: in a variable world, different styles can be favoured under different conditions. Fast explorers may do well when competition is intense or when food patches are ephemeral. Slow explorers may do well when the challenge is to discriminate among similar options where mistakes are costly.

That idea is familiar in animal behaviour, where "personalities" or behavioural syndromes—boldness, caution, aggressiveness—can persist because no single style dominates in all environments. Seeing an analogue in a slime mould is not a reason to attribute personality in the human sense, but it is a reason to treat the organism as something more than a deterministic machine. The mould's distributed physiology can support multiple operating regimes, and those regimes have predictable consequences.

What might a "fast" regime look like in the body of *Physarum*? It can be as simple as reducing the time spent sampling. The organism may send out fewer exploratory fronts or retract them sooner, committing earlier to whichever food source produces the first strong local effect. A "slow" regime might involve maintaining multiple fronts for longer, allowing competing routes to stay alive while information accumulates. Even without invoking any internal variable called "caution," the pattern can emerge from differences in the propensity to expand, in the sensitivity of contraction rhythms to local cues, or in how quickly tubes are pruned when they are not immediately profitable.

The environment can impose urgency in more than one way. Light is a useful laboratory stressor because it simultaneously adds risk and disturbs the organism's mechanics. Hungry moulds also face urgency because their internal reserves are low. Latty and Beekman (2011) manipulated such stressors and found that the relationship between difficulty, time, and error is not a simple straight line. Under harder discrimination tasks, decision times increase; under stress, the mould tends to compress those times. What changes is not merely *how much* it explores but *how long it keeps alternatives alive* before one path begins to monopolise flow.

The word "monopolise" is not metaphorical. In *Physarum*, flow is the great persuader. Cytoplasm streaming through the network delivers nutrients and signalling molecules. A route that begins to draw more flow becomes more robust, because it is better supplied. Competing routes, starved of flow, become weaker and are more likely to be abandoned. This creates positive feedback: early advantages can snowball. Under low urgency, it can be beneficial to delay the onset of this snowball effect, preventing a premature lock-in to a mediocre option. Under high urgency, delaying lock-in may be too expensive, and the organism accepts the possibility of a wrong commitment.

Here is a small thought experiment, close to what happens in the dish. Suppose two food sources are nearly identical, and one is marginally better. The slime mould sends out two fans that reach the patches at similar times. If it commits immediately, random fluctuations—a slightly faster contact, a slightly thicker initial tube—can decide the winner, and the mould might pick the worse option about half the time. If it waits, continuing to feed a little from both while keeping both tubes functional, the marginally better patch will, over time, produce a slightly stronger physiological signal. That signal might be a slightly higher contraction frequency or a slightly greater nutrient flux. Given enough time, the small difference becomes reliable. Waiting is how weak evidence becomes strong.

That logic also explains why mistakes become more likely when the mould decides quickly. Under speed pressure, it cannot average out the noise. It commits on a snapshot rather than on a time-integrated estimate. In statistical terms, it reduces its sample size. In biological terms, it reduces the time available for the network to "feel" the difference between options.

The human parallel is not merely rhetorical. Many of the same mathematical forms that describe evidence accumulation in brains can, in principle, describe evidence accumulation in bodies. A drift toward one option corresponds to a gradual bias in growth or flow. Random fluctuations correspond to local irregularities in tube diameter, contractile phase, or microenvironmental variation. The threshold corresponds to a point at which one route becomes so dominant that the others collapse. The striking thing is not that a slime mould secretly uses equations, but that the equations capture a general feature of adaptive systems: when information is noisy and time is limited, there is no free lunch.

The organism's costs, however, are not the same as ours. A person can decide quickly and then correct the mistake with a sentence: "Actually, I meant the other one." A slime mould corrects mistakes by remodeling its network, and network remodeling is slow. This makes the trade-off sharper. A hasty commitment can lock the organism into a suboptimal architecture that persists long after the initial conditions have changed. One might expect natural selection to favour caution. Yet *Physarum* persists, which suggests that in its world, opportunities often demand speed. Food patches can be transient. Microclimates can shift. A slug may arrive and consume the patch. The mould that hesitates too long may lose everything without ever making an error.

The lab dish also contains a subtlety that real forests amplify: the mould is not choosing once. It is choosing continuously. Even after a route becomes dominant, other parts of the network may continue to probe. A tube that has been pruned can be regrown. In that sense, what looks like an "error" in a short experiment may be, in the longer run, part of a strategy that values rapid exploitation with occasional correction. Human decision-making shows a similar pattern: we often act quickly, then gather more information and adjust, rather than waiting for certainty before moving at all. What differs is the medium of adjustment. For *Physarum*, adjustment is not a revised belief but a rearranged body.

That is why the organism's individual differences are so intriguing. A fast explorer may frequently get the best patch first simply because it arrives early and commits. A slow explorer may arrive later but choose more wisely among what is available. In some contexts, speed creates a competitive advantage that can outweigh occasional errors. In others, careful discrimination saves energy and increases long-term returns. Dussutour and Ma's work (2019) hints that the mould's variability is not noise to be averaged away but a resource for coping with a changeable world.

A final image keeps the trade-off grounded. Picture the slime mould again in a dish with multiple food sources. Under comfortable conditions, it spreads like a cautious investor: small exploratory stakes in several directions, a reluctance to put everything into one option too soon, steady reinforcement of the best-performing route. Under hunger or harsh light, the same organism behaves more like a debtor in a hurry: it concentrates its resources quickly, sometimes backing the wrong option, because the penalty for delay is immediate. The difference is not a change of mind, but a change in how readily the organism lets positive feedback take over.

From the outside, it can look like temperament. From the inside, it is the interplay of energy, stress, and the physics of flow. A system that moves material through a network must constantly decide where to send that material. When time is plentiful, it can afford to distribute broadly and learn. When time is scarce, it chooses a direction and commits. The mould's body makes those policies visible, because every choice leaves a trace in the thickness of a tube, the persistence of a branch, and the pattern of growth that remains after the meal has been taken.

5.3 Rhythms of Consensus

Consensus is usually imagined as a conversation: voices arguing, evidence offered, minds changing. In *Physarum polycephalum*, consensus is a pulse. The organism has no neurons to broadcast a verdict, yet it routinely behaves as a coherent whole. After a period of probing, it stops spreading evenly and begins to *commit*: the bulk of the cell flows into one branch of its network, one corridor of a maze, one food patch among several. The mystery is not that a slime mould can move; it is that a sprawling, many-fronted sheet of cytoplasm can come to agreement about where "it" is going.

The clue lies in a rhythm that is easy to miss until you know to look for it. A healthy plasmodium is never still. Under a microscope, even a strand that seems motionless breathes with slow contractions. The tubes tighten and relax in waves, driven by actomyosin—the same molecular machinery that makes muscles contract, scaled down and spread across a single cell. These contractions push and pull the cytoplasm in a back-and-forth surge known as *shuttle streaming*. Over tens of seconds to a couple of minutes, the direction of flow reverses. It is as if the cell is kneading itself, pumping its own contents through its vascular network.

These pulses are not just a housekeeping mechanism for transport. They are also a means of *communication*. When one region of the network encounters food, its contraction dynamics change. The local oscillations can speed up or strengthen, and that alteration does not stay local. It propagates through the tubes as pressure waves and chemical signals. The result is a physical argument that the whole organism can feel: this branch is doing better; send more resources here.

A historical anecdote makes the logic vivid. In the nineteenth century, long before slime moulds became fashionable in discussions of "computation," physiologists wrestled with a simpler question: how do organs coordinate without wires? The heart offered a dramatic case. Its beat is rhythmic, self-generated, and synchronised across tissue. Early investigators argued about whether the rhythm was imposed by nerves or produced by the muscle itself. Modern physiology vindicated the latter: networks of excitable cells can synchronise through local coupling and produce a coherent beat. *Physarum* offers a parallel lesson in a different key. Its network is not excitable in the neuronal sense, but it is *oscillatory*, and oscillators can couple. Synchrony can arise without a conductor.

The idea of a coupled oscillator can sound abstract, so it helps to anchor it in a concrete laboratory scene. Picture a plasmodium stretched as a branching network on agar, with two food sources at different locations. The organism has already built tubes connecting its body to each source. If you watch carefully, the tubes do not contract randomly. You see waves: a tightening that travels along a strand, followed by relaxation, followed by tightening again. Now place fresh food near one branch. Within minutes, the rhythm near the food changes. The contractions in that neighbourhood become more frequent. In *Physarum*, frequency is not a trivial detail; it is a lever. Different frequencies interacting across a network create gradients in pressure and flow. The branch that oscillates faster begins to influence the direction and magnitude of streaming throughout the organism.

This is where "decision" becomes a mechanical process. A branch that is locally excited by food develops a faster oscillation. Because the tubes are connected, that frequency difference cannot remain private. Pressure fluctuations generated by the faster oscillator travel into neighbouring tubes. The slower regions are not passive; they respond, sometimes adjusting their own rhythms. Over time, the network can settle into a coordinated pattern in which large regions pulse in a coherent relationship. Once that coherence exists, transport becomes biased. Cytoplasm spends more time moving in one direction than the other, or the forward surges become stronger than the backward ones. The organism is now being physically pulled toward the food, not by a single tug, but by a sustained, network-wide asymmetry in its pumping.

One can tell this story at three scales, each adding a piece.

At the smallest scale are the molecular oscillators inside the cell cortex. Smith and Saldana (1992) proposed one influential model in which calcium ions (Ca^{2+}) and phosphorylation cycles in the contractile machinery produce a relaxation-type oscillation. "Relaxation" here means a slow build-up followed by a rapid release, like a toilet cistern refilling and flushing. The details of such models can be technical, but the punchline is simple: the chemistry of contraction is capable of generating a repeating rhythm on its own, and that rhythm can couple to tension and ATP use. The cell does not need an external clock.

At the intermediate scale is the tube. A tube is not just a pipe; it is a muscular pipe. Contraction waves can travel along it, pushing fluid much as peristalsis moves food through an intestine. In a single tube, the relationship between wave pattern and flow is relatively direct. In a network of tubes, it becomes richer. Alim and colleagues (2013) showed that in network-like geometries, the organisation of peristaltic waves can maximise transport, and that the wavelength of contraction patterns tends to scale with the size of the organism. This is a remarkable kind of self-tuning: the *beat* is not fixed but adapts as the network grows, as if the organism were adjusting its internal logistics to match its own changing dimensions.

At the largest scale is the whole plasmodium as a distributed system. Here, global oscillations can be thought of as dominant modes of a soft, contractile network—patterns that the entire structure can support robustly. The organism does not have to compute these modes explicitly. By virtue of being an elastic, fluid-filled network with active contractility, certain patterns are easier to sustain than others. The environment then acts as a selector. Food stabilises certain modes by feeding the branches that support them; stressors destabilise others by weakening contraction. Consensus is, in part, the network settling into the mode that best fits current conditions.

A concrete example of this settling process comes from confinement and harsh illumination. In some experiments, *Physarum* placed in stressful, light-limited environments does not immediately find an efficient way out. It cycles through different dominant contraction patterns, almost as if testing different internal "gears," before settling into the pattern that produces the most effective transport and coincides with escape. The important point is not the drama of being "trapped" but the mechanism implied: the organism can switch between global dynamical states, and those switches matter for behaviour. A choice, in this view, is a change in the global rhythm of pumping.

Why does frequency matter so much? Because flow is not just a consequence of contraction; it is also a means of integration. The cytoplasm carries nutrients and signalling molecules throughout the organism. If one branch is feeding well, the products of that feeding—energy, metabolites, perhaps chemoattractant uptake and internal signals—are carried into the rest of the network by streaming. This alters the internal state of distant regions, changing how readily they contract and how strongly they grow. Information is thus embodied in circulation. A brain sends messages as spikes; *Physarum* sends messages as altered flow.

A useful way to see the logic is to imagine two arms of the organism competing. Each arm has a local oscillator with frequency f_1 or f_2. Suppose arm 1 is near high-quality food and increases its frequency ($f_1 > f_2$). The faster oscillation generates pressure pulses more often. When coupled to the rest of the network, this can bias net flow so that, over time, more cytoplasm is delivered to arm 1. That delivered cytoplasm contains more contractile machinery and more capacity to thicken tubes. Arm 1 becomes even better at pumping and feeding. Arm 2, receiving less, gradually loses strength. The system amplifies a small frequency difference into a large structural difference. That is how a distributed organism "votes" without counting ballots.

If this sounds too smooth, it should. Real plasmodia are messy. Their oscillations can be noisy. Their network topology changes as they explore. Yet noise does not destroy the principle. In fact, noise can be useful, because it allows exploration of alternatives. The earlier dilemmas—risk versus reward, speed versus accuracy—make sense only if the organism can, at least for a while, keep multiple options alive. Oscillatory dynamics provide a natural way to do this. When coupled oscillators are near a tipping point, small perturbations can shift dominance. A slight advantage in one branch can become decisive, but only after a period of competition. That period is the organism's deliberation, expressed as a fluctuating pattern of streaming and contraction.

There is also another layer of distributed intelligence that does not sit inside the body at all. Reid and Latty (2012) showed that *Physarum* avoids areas it has previously explored, effectively using its own slime trail as an *externalised spatial memory*. This matters for consensus because it changes what the organism samples. A brain can remember "I tried that corridor." The slime mould writes the memory into the environment. When it later approaches the same area, chemical cues from the deposited slime bias its growth away. The environment becomes part of the decision-making loop.

This external memory has a particular emotional resonance because it echoes a very human way of coping with complexity: leaving notes. We put a sticky note on a door we should not forget to lock. We mark a page in a book. We drop a pin on a map. The slime mould does something analogous, except the note is secreted mucus and the reader is the same cell that wrote it. In trap-like navigation tasks—

for example, a U-shaped barrier that draws an agent into a dead end if it simply follows the scent of food—this external memory can make the difference between endless dithering and successful escape. By avoiding its own trail, the organism prevents itself from repeatedly wasting effort in the same unproductive region. Consensus becomes easier when the space of options is pruned, and pruning can be outsourced.

The interplay between internal rhythm and external memory suggests a general principle. Decisions in *Physarum* are not made at a single point or instant. They are enacted across time by the coupling of oscillations and the redistribution of flow, and across space by the marks left in the environment. What looks like a single organism "choosing" is, mechanistically, a repeated negotiation between local events (a food contact here, a light stress there), global dynamics (which contraction mode dominates), and environmental traces (which regions have already been tried).

One might worry that calling this "consensus" smuggles in politics. Yet the term captures something precise: the organism often begins in a pluralistic state, with multiple growing fronts, and ends in a monolithic state, with a dominant route. The transition from pluralism to unity is mediated by coupling. In a parliament, coupling is speech and persuasion; in *Physarum*, coupling is pressure, flow, and chemical feedback. Either way, consensus means that local differences have been reconciled into a single coordinated action.

The diagram below condenses the mechanism into a visual shorthand: tighter wave patterns represent faster contraction near food, and the arrows represent the resulting net flow of cytoplasm toward the better option. The key idea is that a frequency difference is not merely a measurement; it is a force that rearranges the organism.

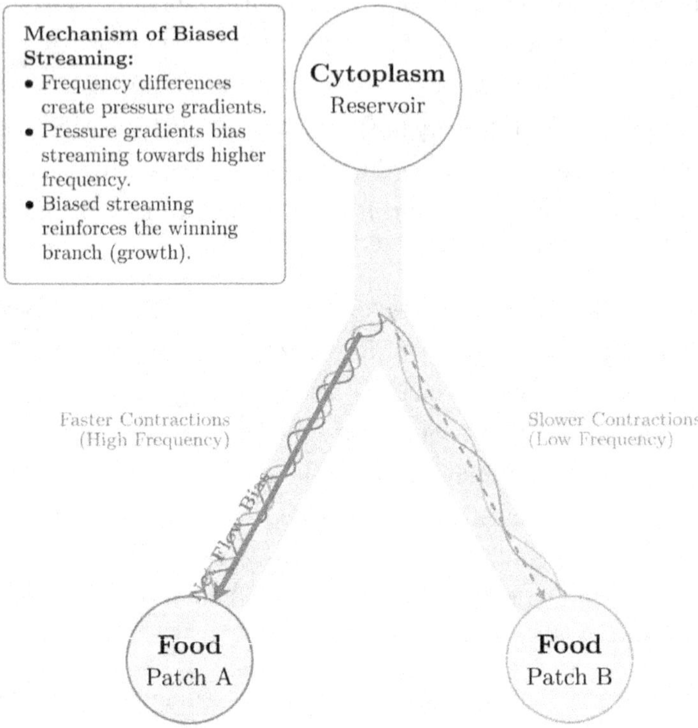

Once you see the organism as a network of coupled pumps, several earlier puzzles become less mysterious. Taking a risk for a better reward can be understood as allowing a stressed branch to compete longer before it is pruned, or as raising the reward signal enough to overcome the damping effect of danger on oscillations. Deciding quickly under hunger can be understood as lowering the level of coherence needed before flow becomes strongly biased, letting the system lock into a dominant mode sooner. Individual differences can be understood as different parameter settings of the same machinery: how readily contraction frequency changes with food, how strongly tubes reinforce with flow, how quickly unused branches decay.

None of this requires a central controller. In fact, a controller would be hard to locate. The closest thing to a "command" is the emergence of a global oscillation pattern that the entire network falls into, like a crowd settling into a chant. The chant is not a message; it is an organising force. It moves fluid, and fluid movement reshapes the body. That is why the slime mould's decisions are so hard to separate from its physiology: the mechanism of consensus is the mechanism of staying alive.

It is also why watching a plasmodium can feel oddly intimate. Under the microscope, the pulses are slow enough to match a human sense of time. One can sit with the organism and see its internal debate become a steady rhythm, see hesitation become commitment, see a distributed sheet of life pull itself together and begin to go, not with a twitch but with a deep, deliberate beat. The verdict is not spoken. It is pumped.

Chapter 6

Memories Written in Slime

Navigating a complex world typically requires a map stored within a brain, yet the brainless slime mold masters mazes with uncanny efficiency. By offloading memory onto the physical environment through chemical trails, Physarum polycephalum transforms its surroundings into an external hard drive, allowing it to escape traps that often baffle sophisticated algorithms. This chapter reveals how the organism reads its own history etched into the substrate to guide future decisions.

6.1 The Hansel and Gretel Strategy

In the Grimm story, Hansel's trick is simple: he makes the forest remember him. Each pebble is a small refusal to be lost, a way of turning a confusing space into a readable record of where he has already been. A slime mould performs a version of the same maneuver, but with a twist that is easy to miss if you imagine intelligence as something that happens only inside a head. *Physarum polycephalum* writes its own history into the world as it moves, and then consults that history later. The medium is not a trail of stones but a thin, extracellular slime sheath: a non-living layer of secreted material left behind on the surface it traverses. The outcome is not nostalgia but navigation.

Calling this a *memory* risks sounding like poetic license. Memory, after all, is supposed to be stored *in* an organism, in some internal substrate that can be accessed at will. Yet what matters in practice is function: past experience changes future action. The slime sheath does exactly that, not by encoding a story, but by biasing the organism's next move away from its own footprints. The environment becomes a notepad that is written automatically, without deliberation, and read continuously, without a librarian. The trick is particularly revealing because it solves a kind of problem that defeats many simple machines: the trap that looks like a corridor.

Imagine a small robot designed with the most straightforward rule for finding a target: move in the direction where the signal is stronger. The signal might be light, heat, or a chemical plume. This strategy, called *gradient following*, works beautifully in open spaces. But it can fail catastrophically in environments with obstacles. One classic demonstration is the U-shaped trap. Place the robot outside a U-shaped barrier, and put the target (the attractant) inside the U, near the back. The signal leaks out of the opening and creates a gradient that points into the trap. The robot obediently enters, and once inside it continues to chase the increasing signal, pressing deeper into the U. When it reaches the dead end, the strongest signal is still *forward* (straight through the wall it cannot cross), so the robot dithers: it turns a little, tries again, turns back, tries again. Without a mechanism to register that it has already explored the dead-end boundary, it can become stuck in a loop of sincere but futile effort.

A slime mould is also attracted by chemical cues from food. Put an oat flake or another nutrient source inside the U-shaped barrier and the organism, sensing the gradient, spreads into the trap. At first glance it seems destined for the same fate as the robot: the strongest cue is in the dead end, and the barrier is impenetrable. But *Physarum* has an extra ingredient, literally: it leaves behind a chemical trace that it later treats as a warning label. As it explores, it coats the surface with slime. Later, when it encounters that slimed region again, it behaves as if the substrate has become less desirable. The effect is not absolute repulsion, like a brick wall; it is more like a negative vote in a decision that is being tallied continuously across the organism's body. Over time, these negative votes accumulate along repeatedly visited boundaries, and the organism is pushed away from the dead end toward the opening it came through. What looks like a trap becomes a temporary detour.

The decisive point is that the slime trail is not a mystical aura, and it is not a map stored in an internal brain. It is matter deposited outside the body. The sheath persists long enough to matter on the timescale of exploration. That persistence is crucial. Hansel's breadcrumbs fail because birds can erase them. A slime mould's trail is not so easily eaten by chance; it is a physical modification of the surface, a chemical memory written into the world.

This behaviour was made crisp in experiments that posed the U-shaped problem explicitly and then altered whether the organism could make use of its own traces. When researchers such as Chris Reid and Tanya Latty (2012) prevented *Physarum* from accessing the information in its previously laid slime—for example, by manipulating the substrate so the trail cue was masked—the organism's performance dropped. When the trail was available, escape was more reliable. That comparison matters because it separates two possibilities that are otherwise hard to disentangle. One possibility is that the slime mould simply integrates the current chemical gradient and eventually stumbles out by chance. The other is that it uses history-dependent information. The U-shaped trap makes the difference visible: pure gradient following tends to stick; gradient following plus a memory of where you have already pressed tends to release.

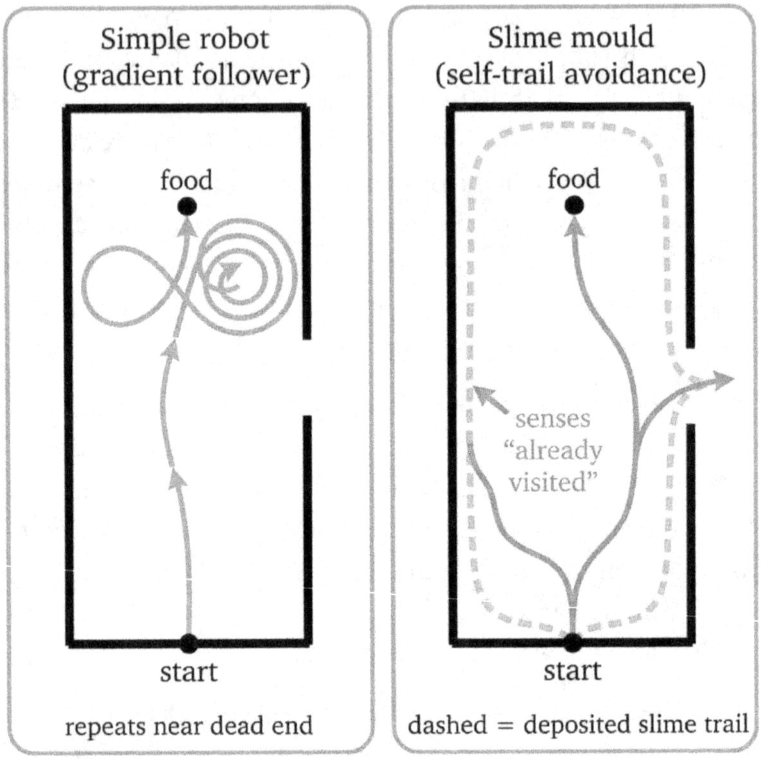

To understand why this works, it helps to picture the organism not as a blob that decides once, but as a living sheet of coupled local decisions. The plasmodium—the multinucleate, single-celled stage of *Physarum* that is used in most experiments—extends pseudopod-like fans and thin veins across a surface. At the front edge, the organism samples chemical cues: nutrients, repellents, moisture, perhaps even subtle texture. At the same time, it secretes slime as it glides. Hours later, that same edge (or another part of the body) may encounter the very substrate it once covered. The slime sheath is then part of the sensory environment. It is not *inside* the organism, but it is coupled to it in a tight loop: movement produces the trace; the trace reshapes movement.

This is a spatial memory of a particular kind: an exclusion memory. It does not say, "Food was here," so much as "I have already been here." In computational language, it is a record of visited states. Many search algorithms become dramatically more efficient if they can mark explored nodes and avoid revisiting them. The familiar puzzle of solving a maze by keeping one hand on a wall is a crude way of using the environment as memory; marking corridors with chalk is a better one. In the natural world, insects do something similar. Ants lay pheromone trails that recruit nestmates, and some species also deposit "no entry" signals on exhausted food patches. The slime mould's version is solitary rather than social, but the logic is the same: change the world in a way that makes future action cheaper.

There is a subtle but important distinction between using slime as a mechanical aid and using it as information. Many organisms secrete mucus or slime to reduce friction, to prevent desiccation, or to lubricate movement. *Physarum* likely benefits mechanically as well. But the behavioural experiments suggest that the slime sheath is not merely grease: it is discriminated. In Y-maze assays, where the organism faces a fork and must choose between two arms, *Physarum* tends to avoid an arm that has been pre-coated with its own slime compared with a clean arm. Even more intriguingly, it can show different responses depending on who laid the trail: a conspecific versus a different species. That kind of specificity is hard to explain as a simple mechanical effect. It implies chemical signatures that the organism can detect and treat differently.

The Hansel-and-Gretel analogy is tempting precisely because it captures both the brilliance and the fragility of the strategy. The trace must persist long enough to be useful, but not so long that it becomes a permanent constraint. In a changing environment, yesterday's "already visited" might contain today's new opportunity. A slime mould foraging on a forest floor would be foolish to treat its own trail as a sacred taboo, because the world is not a static maze drawn in plastic. Leaves decay, microbes bloom, rain redistributes nutrients. The memory in slime must therefore be *soft*, a bias rather than a rule.

That softness shows up in the way avoidance interacts with attraction. When food cues are strong, *Physarum* can override its reluctance and cross a slimed region. This makes biological sense. A rigid prohibition would strand the organism if the only route to a rich patch required crossing its own past. Instead, the trail acts like a cost added to the decision, not an unbreakable wall. The organism is performing something close to an economic calculation, but without numbers: the benefit of chasing the nutrient gradient must outweigh the cost of re-entering explored territory. The cost is provided by the trail; the benefit by the food.

One way to feel the plausibility of this without any equations is to imagine the organism as a distributed committee where each point along the edge votes on where to grow next. Food increases the enthusiasm to advance; slime decreases it. If every direction has some slime, the least slimed direction wins by default. If one direction has a powerful food cue, it can win even if it is slimed. The result is a flexible search that neither wanders aimlessly nor becomes trapped by its own history.

A concrete historical moment helps anchor this. In 2000, Toshiyuki Nakagaki and colleagues popularized a striking demonstration: place oat flakes at two locations in a maze and a piece of *Physarum* at one food source. The organism spreads through the corridors in many directions, then gradually retracts and leaves behind a thickened path that connects the two food sources along an efficient route. The image of a slime mould "solving a maze" became famous because it seemed to show planning. But the more interesting story is about *how* the maze becomes solvable when you have a body that can occupy many corridors at once and a way of pruning. Trail avoidance is not the only mechanism in play there—flows through tubes reinforce some branches while others decay—but it fits the same general theme: history is not stored as a blueprint; it is embodied in the world the organism has already modified. The maze is not solved by an internal overview; it is solved by a succession of local commitments that become harder to undo as the organism invests in them.

The U-shaped trap isolates the trail mechanism even more cleanly than a maze does because it strips away the romance of "finding the shortest path" and focuses on a grimly practical question: how do you avoid wasting time in a cul-de-sac? In robotics, engineers address this with mapping, with memory registers, with algorithms that explicitly store visited locations. But those solutions come with overhead. Mapping requires sensors, computation, and a coordinate system that remains stable as the robot moves. A slime mould does not need any of that. It turns the ground itself into the coordinate system: a place is "the same place" because it still bears the mark of having been touched. The memory is *indexed by location* automatically, because it is located.

This is why the slime sheath is such a clean illustration of distributed intelligence. It is not intelligence in the sense of symbol manipulation. It is intelligence in the sense of reducing future error by recording past action. It is also intelligence in the sense of humility: rather than insisting on keeping everything inside, the organism offloads. Offloading is not cheating; it is what makes complex behaviour possible with minimal internal machinery. Humans do it constantly. We write shopping lists. We leave books on the stairs as reminders. We arrange objects in a room so that forgetting becomes harder. The slime mould's list is not written in language, but the logic is recognizable.

There is a deeper implication, and it concerns how we decide what counts as a cognitive boundary. If the slime trail changes behaviour in a way that depends on past action, then the relevant "memory system" is not confined to the living membrane. It is a coupled organism—environment circuit. This does not mean the trail is magic. It does mean that the unit of analysis shifts. When we ask, "How does *Physarum* remember where it has been?", a complete answer must include the chemistry of secretion, the persistence of the sheath, the sensory machinery that detects it, and the motor dynamics that translate that detection into turning away. Remove any one of these, and the memory dissolves. The trail is not merely an output; it is part of the loop that produces the next input.

It is also, in a quiet way, an argument against a certain kind of technological arrogance. Engineers often treat the world as a nuisance: noise, friction, uncertainty. The desire is to build machines that can ignore messy environments by imposing internal order. The slime mould does the opposite. It embraces the environment as a workspace, and it makes that workspace more informative as it goes. It is not that the organism is cleverer than a robot; it is that it recruits more of the world into the computation.

In practice, this means that a slime mould can behave like an explorer who never needs to stop and consult a map, because the act of walking leaves behind a faint chalk mark that says, "Not here again unless you must." It is hard to imagine a more modest form of memory, and hard to imagine a more effective one for an organism whose central problem is not contemplation but finding dinner before drying out. The forest does not become friendly, but it becomes legible, traced with a private script written in slime.

6.2 The Calculus of Exploration

A hungry organism has a deceptively mathematical problem. Stay close to what you already know and you eat reliably, but you may miss something better. Range farther and you might find a feast, but you might also waste time, energy, and moisture on emptiness. This tension is often called the *exploration—exploitation trade-off*: exploration means sampling new territory; exploitation means returning to, and investing in, what has already paid off.

Humans meet the same dilemma constantly. Do you order your usual at a restaurant, or try the unfamiliar dish? Do you keep your savings in a dependable account, or take a risk on an investment that could either grow or evaporate? What makes the dilemma feel genuinely cognitive is that it is not solved by a single reflex. It is solved by *adjusting* a bias as conditions change: when you are desperate, you may gamble; when you are comfortable, you may conserve. The surprise, and the charm, in *Physarum polycephalum* is that it carries out an analogous adjustment with no central executive to weigh pros and cons. The calculus is written into its movements, its flows, and its willingness to override its own reluctance.

The previous idea to keep in mind is that *Physarum* leaves behind information on the surface it traverses: a slimy residue that makes previously visited ground less attractive. That residue is not a prison wall; it is a negative bias. Once you see it that way, the exploration—exploitation problem becomes vivid. If the organism treated its trails as forbidden territory, it would be a poor forager in a world where food patches are sparse and routes must often cross. If it ignored its trails entirely, it would waste time in loops and dead ends. A useful strategy lies in between: avoid your own past when you can, but be willing to revisit it when the reward is worth the cost.

This in-between strategy is not merely a philosophical stance. It appears as a measurable behavioural rule in laboratory assays. In a simple Y-shaped maze, where the organism must choose between two identical arms, it tends to prefer the arm with clean substrate over an arm that has been coated with its own slime. Change the situation by placing food in the slimed arm, and the choice can flip. That flip is the signature of a trade-off. The trail acts like a faint *tax* on revisiting. Food acts like income. Depending on the balance, the organism either pays the tax or walks away.

To speak of taxes and income is of course metaphor. What is not metaphor is the physical basis for the balance. The organism senses chemical cues from nutrients (*chemotaxis*, meaning movement guided by chemicals). It also senses the chemical and perhaps mechanical consequences of its own deposition. The front edge is continually sampling. But the organism does not move like a wheeled vehicle that points one way and goes. It is a living network whose internal flow reorganizes itself. If a branch begins to carry more cytoplasmic flow, it tends to thicken and persist; if it carries little flow, it thins and may disappear. The outcome is a shifting investment portfolio: resource is allocated to some branches and withdrawn from others.

A historical anecdote helps, because the field has a habit of appearing magical until one learns what was actually done. In the late 2000s, Atsushi Tero and colleagues (2010) became famous for showing that when *Physarum* is offered multiple food sources in a dish, it often builds transport networks that resemble human-designed networks in their efficiency and redundancy. One celebrated comparison involved the rail system around Tokyo: given oat flakes arranged roughly like cities, the organism formed a web of tubes that balanced short routes with backup connections, not unlike a commuter map that has grown under decades of constraints. It was an irresistible headline: the slime mould "designs" railways. The more sober lesson is deeper. A network can be optimized without a planner if local reinforcement rules couple to global flows. In other words, the organism can solve a large-scale allocation problem because it *moves resource through its own body* and lets the pattern of movement decide where more body should exist.

Exploration and exploitation are the two faces of this same allocation. Exploration requires many thin probes outward, a willingness to spend material on possibilities. Exploitation requires consolidation: thick, reliable routes between known rewards. The organism's behaviour often shows an initial expansion phase, during which it spreads broadly, followed by a pruning phase, during which it retracts from unprofitable areas and strengthens the profitable ones. That change of phase can look like a change of mind, but it can also be understood as the natural consequence of competing tendencies: growth at the frontier and reinforcement in the interior.

The classic maze demonstration by Nakagaki and colleagues (2000) makes this visible in a way that is hard to forget. Place food at two points in a maze and introduce *Physarum* at one food source. The organism spreads into many corridors, sampling them in parallel. That is exploration at its most literal: the plasmodium pours itself into the available space. Over time, however, the organism retracts from many corridors and leaves a dominant tube connecting the two foods. That is exploitation: rather than maintaining a large, costly footprint, it concentrates on the route that reliably moves nutrients between valuable points.

It is tempting to treat this as a two-step program: first explore, then exploit. But the more faithful picture is a continuous competition, a running tally in which different parts of the organism are constantly being compared. A branch that happens to reach food begins to experience stronger chemotactic signals, and it may transport more nutrient-laden cytoplasm. Higher flow can reinforce the tube, making it larger and less resistive, which attracts even more flow. This positive feedback is exploitation made physical: the organism commits more of itself to what is already working. Meanwhile, branches that do not deliver reward may still expand for a while, but without reinforcement they tend to be starved of flow and eventually shrink. Exploration is not cancelled; it is outcompeted.

At the same time, the external trail bias acts as a kind of *regularizer* on exploration. A purely expanding organism could waste time repeatedly probing the same areas. By making revisitation slightly unattractive, the slime sheath encourages the frontier to push into genuinely new territory. A self-avoiding walk on a graph is a familiar mathematical object: a path that tries not to step on its own footsteps. It covers space more efficiently than an unguided random walk, which tends to tangle around itself. *Physarum* is not executing a formal algorithm, but the resemblance is instructive. The slime trail turns the foraging process into something closer to a self-avoiding search, while the ability to override the avoidance prevents the search from becoming brittle.

This brittleness matters because a world with obstacles can punish strict rules. Suppose a food patch lies across a narrow bridge of ground. If the organism crosses once, the bridge becomes slimed. If slime avoidance were absolute, the organism would be unable to cross again, cutting itself off from a resource it has already discovered. A sensible strategy is to treat the slimed bridge as less attractive than fresh ground, but still crossable if the alternative is worse. Many experiments point in this direction: trail avoidance is strong enough to steer the organism away from needless loops, but weak enough to be outweighed by a strong attractant or by necessity.

The word *calculus* is often used to suggest something grand and explicit: symbolic manipulation, deliberate optimization, equations on a blackboard. Here it means something humbler and closer to its original sense of “a method for reckoning”. *Physarum* reckons with costs and benefits using gradients, flows, and decay. The cost of maintaining tissue is real: tubes require cytoplasm, and exposed surface risks drying. The benefit of maintaining a tube is also real: it provides a low-resistance route for transporting nutrients and signals. These two quantities are not written down; they are enacted.

Theoretical work grew alongside the experiments because the behaviour seemed to invite formalization. Tero and Kobayashi (2006) developed a model sometimes called the *Physarum solver*, in which edges of a graph represent possible corridors and their “conductivity” increases with flow and decreases otherwise. Run the dynamics, and the network tends to converge on efficient paths. Later theoretical results, including work by Bonifaci and colleagues (2013), showed that in certain settings such reinforcement-and-decay dynamics can be proven to find shortest paths. These proofs are not needed to appreciate the organism, but they sharpen the intuition: under some reasonable rules, a distributed system that reinforces high-flow routes and lets low-flow routes wither will, in effect, compute an optimal or near-optimal transport plan.

One can now see how the exploration—exploitation dilemma is embedded in the organism’s architecture. If you reinforce too aggressively, you exploit too early. You commit to a route before you have sampled alternatives, and you may miss a better option. If you reinforce too weakly, you never exploit, and you remain a diffuse mess of tentative branches. Somewhere in the dynamics is a sweet spot where the organism explores enough to find good opportunities, then consolidates enough to profit from them.

A useful concrete image is to imagine the organism as a city deciding which roads to pave. In the beginning, there are dirt tracks everywhere, because people try many routes. As traffic concentrates on some routes, those routes are paved and widened, which attracts even more traffic. Other routes fall into disuse and fade. The difference is that in the slime mould the traffic is cytoplasm, the paving is thickened tube wall, and the city council is nowhere to be found. The decisions are made by feedback between flow and structure.

But unlike a city, *Physarum* is also a single, continuous body. If one region is drawing heavy flow, other regions may be deprived. That constraint introduces a finite-budget problem: investment in one branch limits investment elsewhere. The tug-of-war metaphor used in some bio-inspired algorithms captures this neatly. When resources are conserved and redistributed, competing options pull against each other. A branch that begins to win can draw more resource, making it win harder. Yet the system can still remain sensitive to new information if the redistribution is not too rigid. In such a regime, a surprising new food signal can still steal resources back from an entrenched route.

There is a human resonance here, because the same tension governs our own habits. A habit is exploitation: it saves energy by reusing a known solution. Curiosity is exploration: it spends energy on uncertainty. Too much habit and life shrinks; too much curiosity and nothing is finished. In *Physarum*, the analogue of habit is the reinforced tube, and the analogue of curiosity is the continual probing of the frontier. The slime trail adds a subtle twist: it is as if the organism leaves itself sticky notes saying "you've tried that already," but the notes can be ignored in an emergency.

One can even glimpse why such a system might be evolutionarily stable. A strict self-avoider would be excellent at covering new ground, but poor at exploiting a discovered patch that requires revisiting a route. A non-avoider would repeatedly waste effort, especially in cluttered environments, and would be vulnerable to traps. A flexible avoider can do both, and flexibility is often what selection rewards in variable habitats.

The story becomes more poignant when one remembers the scale at which all this happens. The plasmodium is not a large animal striding across terrain. It is a thin living film on a damp surface. Dryness is a constant threat. Time is measured in hours. When it allocates itself into a dead end, it is not merely inconvenienced; it is exposed. When it retracts from an unprofitable corridor, it is not merely changing strategy; it is reclaiming water and cytoplasm. Its calculus is a calculus of survival.

Some of the most illuminating moments in the laboratory come when the organism is given a choice that forces an explicit trade. Put two food sources at different distances: one close and modest, one farther and rich. Or arrange patches so that reaching the richer one requires crossing an area already coated with slime. The organism's behaviour can shift with hunger, with humidity, with the strength of food cues, and with the costs of travel. These are not mood swings. They are the signs of a system whose decision thresholds are adjustable, tuned by physiological state and environmental conditions.

Engineers have taken notice for practical reasons. In robotics and computer science, exploration—exploitation dilemmas show up in path planning, in network routing, and in machine learning systems that must decide between trying new options and sticking with known good ones. Many of these systems solve the problem by storing explicit values and updating them. *Physarum* suggests another route: let the world and the body jointly store the values. A trail becomes a negative value stamped onto space. A reinforced tube becomes a positive value stamped onto structure. Flow becomes the currency that converts those stamped values into action.

The result is not perfect rationality. It is something arguably more interesting: a rationality that is inseparable from material. When *Physarum* withdraws from a region, it does not merely stop going there; it often dismantles the tube itself, recycling material. When it commits to a corridor, it does not merely prefer it; it reshapes itself so that preference becomes easier to enact next time. The calculus is therefore not only about where to go, but about what to *be*. A branch that is allowed to persist is a branch that becomes, in a literal sense, part of the organism's future options.

This perspective can make even familiar human experiences look slightly different. We often imagine decision-making as something that happens first, and then is implemented by action. In *Physarum*, decision and implementation are entangled. The organism decides by changing its own conductances, and it changes its conductances by moving. It learns by leaving traces, and it leaves traces by learning. The trade-off between exploring and exploiting is not managed by a supervisor; it emerges from a conversation between a moving edge, a secreting surface, and a pulsing internal flow that continually reassigns importance to different possibilities.

A slime mould on a Petri dish, spreading and retracting, can look slow enough to be tranquil. Underneath that calm pace is a relentless accounting: each branch is being charged rent, each new probe is an investment, each reinforced tube is a bet that the future will resemble the recent past. The organism cannot know that future, but it can arrange to be less wrong. In that modest ambition—to be less wrong at the next step than at the last—the calculus of exploration becomes a kind of quiet intelligence, one that never announces itself, but keeps the body fed.

6.3 Externalizing the Mind

Memory is usually imagined as a private possession. You carry it around; you lose it when you sleep too little; you fear it when it begins to falter. Even when we talk about "collective memory" we often mean something metaphorical, a shared story rather than a material record. Yet much of human thinking is not private at all. We lean on calendars, shopping lists, street signs, and the choreography of familiar rooms. A child learns to tie their shoes not only by rehearsing movements in their head, but by using the lace itself as a tactile guide that constrains what happens next. A mathematician spreads pages across a desk because the arrangement becomes part of the proof. The mind, in these cases, is not a sealed box; it is a system that includes hands, tools, and a world that has been made easier to think in.

This perspective was sharpened into a philosophical proposal by Andy Clark and David Chalmers in the late 1990s: the *extended mind* thesis. If an external object is reliably available, routinely used, and tightly coupled to a person's activity, then it may function as part of the cognitive process rather than as a mere aid. Their most famous thought experiment involved a notebook used by someone with memory impairment: the notebook is consulted as automatically as biological memory would be, and so, the argument goes, it deserves to be counted as part of the memory system. People can disagree about the metaphysics—about whether notebooks literally become part of the mind—but the functional point is hard to escape. Intelligence often depends on moving information out of the head and into the world.

Slime moulds make this argument without a thought experiment. The organism has no neurons, no synapses, no centralized organ that looks like a command post. And yet it shows behaviours that, in other animals, would usually be explained by appeal to a nervous system: avoiding dead ends, allocating effort between opportunities, changing future action based on past experience. The key manoeuvre has already appeared in concrete form: *Physarum polycephalum* deposits slime as it moves and later responds to that deposited material. In human terms, it writes a note to itself and then reads it. The striking thing is that the note is not written in a special place, like a brain, but right there on the ground.

This is a good place to introduce a term from the study of social insects and self-organizing systems: *stigmergy*. The word was coined by the French biologist Pierre-Paul Grassé in the 1950s while studying termites. He was fascinated by the fact that termites build elaborate nests with arches, pillars, and ventilation systems, despite each individual insect having limited information and limited abilities. Grassé argued that coordination could happen through the work itself: a termite modifies the environment by placing a pellet of mud laced with chemicals, and that modified environment changes what other termites do next. The work product becomes a signal. Communication is not only direct (insect to insect); it is indirect, mediated by traces in the world. A trail of pheromone, a pellet of soil, a partly built wall: these are messages written in matter.

The elegance of stigmergy is that it can replace an internal plan with an external feedback loop. You do not need every worker to know the blueprint if the emerging structure guides the next action. You do not need a leader if the environment, continually updated, functions as a shared memory. In an ant colony, a pheromone trail is often described as a collective external memory: it encodes where food has been found, how recently, and with what level of traffic. It can fade, preventing the system from becoming stuck in yesterday's success. It can be amplified, allowing rapid recruitment. The colony thinks with scent.

A slime mould is not a colony, and it does not coordinate multiple individuals. It is, however, a sprawling system with many semi-independent decision points along its perimeter. Its problem is not how to coordinate workers, but how to coordinate *itself*. When different regions of the plasmodium spread into different corridors, how does the organism avoid becoming a disorganized splatter of conflicting tendencies? Stigmergy offers a lens: coordination can occur through environmental traces (the slime sheath) and through structural traces (reinforced tubes) that constrain what happens next. The key is coupling. The organism changes the world. The changed world changes the organism.

One way to make this less abstract is to consider a humble human example that has nothing to do with insects. Think of the way people manage a crowded kitchen while cooking. A well-run kitchen relies on *place*: knives returned to a specific spot, dirty dishes stacked in a predictable location, ingredients grouped in the order they will be used. Much of the coordination is not in anyone's head; it is in the layout. If the cutting board is cleared, it signals readiness for the next task. If a pot is moved to a back burner, it signals that it no longer needs attention. These are stigmergic cues: traces that guide action without verbal negotiation. Even alone, you use them. You leave the lid ajar to remind yourself that the contents are hot. You place a spoon across a bowl so you do not forget to stir. The environment becomes a prosthetic for attention and memory.

A slime mould does something analogous at a different scale and with different materials. The slime sheath is one such cue. It is produced as a by-product of movement, but it becomes a guide for movement. The organism does not need to maintain a precise internal record of where it has been; the world holds that record as a patchwork of altered substrate. Crucially, this record is spatially indexed without extra effort. Human beings work hard to make maps that preserve spatial relationships. *Physarum* gets that for free: the mark is located where the experience occurred.

This idea can feel unsettling because it scrambles a comforting boundary: the boundary between organism and environment. The slime sheath is not alive, yet it can influence behaviour in ways that look like memory. That suggests that memory, functionally, can be a property of a *system* rather than a property of a *tissue*. If the organism routinely writes information to an external medium and routinely reads it back in a way that influences future choice, then excluding that medium from the cognitive account becomes a matter of preference rather than necessity.

This is not only a philosophical quibble; it changes what one tries to measure. Suppose one is skeptical and insists that the slime sheath is merely a repellent, a kind of dirt the organism dislikes. Even in that minimal description, past action affects future action via an external mechanism. But the experiments that reveal specificity and context-sensitivity push beyond the idea of generic dirt. In Y-mazes, *Physarum* can distinguish between substrates marked by different organisms. In trap experiments, the presence or absence of accessible trails changes the probability of successful navigation. These behaviours suggest that the external trace is not just an irritant; it carries usable information.

The more interesting point, however, is what this implies about neurons. It is easy to slip into an assumption that learning requires synapses, that memory requires circuits, and that decision-making requires something like a brain. Slime moulds undermine that assumption in a polite but persistent way. They show that some functions we associate with brains can be realized by other architectures. A brain is one solution, not the only solution, to the problem of using the past to shape the future.

At this point a necessary nuance comes into view. The slime mould's most vivid memories are written in slime, but not all of its memory-like behaviour can be offloaded to the environment. Researchers have demonstrated forms of learning that look internal. One is *habituation*, a basic phenomenon in animal learning in which an organism's response to a repeated harmless stimulus gradually diminishes. If you hear the same ticking clock long enough, you stop noticing. In

Physarum, habituation has been studied using mildly unpleasant substances that the organism initially avoids. With repeated exposure, the avoidance weakens. Careful studies such as those by Boisseau and colleagues (2016) have addressed a common skeptical retort—"it's just fatigue"—by testing hallmarks used in animal research: the response decreases with repetition, recovers after rest, and shows some degree of specificity rather than collapsing into general exhaustion. Even more strikingly, work by Boussard and colleagues (2019) suggested that habituation-like effects can persist through a dormant stage, meaning that after the organism enters a resting form and later revives, the altered response can remain. There is no slime trail in that case to consult; the memory must be stored in the organism's physiology, perhaps in the state of biochemical networks or in structural properties that survive dormancy.

Another famous example, reported by Saigusa, Tero and colleagues (2008), involves anticipation of periodic events. If the organism is exposed to a stimulus at regular intervals—for instance a change in humidity or temperature—it can begin to show a predictive change in behaviour at the expected time even when the stimulus is omitted. This looks like a primitive form of time-keeping: not a clock that tells time in numbers, but a dynamical system that acquires a rhythm. Again, there is no external mark that obviously encodes the schedule. The memory appears to be a property of internal dynamics.

Why bring these internal forms of memory into a discussion of externalizing the mind? Because they prevent a tidy but misleading conclusion. It would be easy to romanticize the slime trail as if it explains everything, and to claim that the organism has outsourced its mind entirely. The truth is more interesting. *Physarum* seems to exploit multiple memory channels: one that is explicitly external and spatial (the slime sheath), and others that are internal and temporal (habituation, anticipation). The organism is therefore not a counterexample to neural cognition so much as a reminder that cognition is modular in a broad sense. Different problems are solved by different forms of persistence.

Once this is appreciated, the trail itself can be seen less as an isolated trick and more as a component in a larger repertoire of feedback loops. The slime sheath is a way of making space remember. Habituation is a way of making the body remember. Anticipation is a way of making time remember. None of these requires a neuron, but all require a system with state that can be updated.

Stigmergy, in this context, is not merely a label borrowed from termite mounds. It becomes a general principle: the world can be turned into a memory buffer, and that buffer can substitute for internal storage when the coupling is tight enough. For a slime mould, tight coupling is built in. The organism is always in physical contact with the substrate. It is always secreting and sensing. Movement and perception are not separated by a large distance or a long delay; they are intertwined at the leading edge.

There is a quiet lesson here for how intelligence may have evolved, and for how it might be engineered. If you begin with a system that can move and can leave traces, you already have the skeleton of a memory mechanism. Add the ability to sense those traces, and you have the beginnings of self-avoidance and efficient search. Add the ability to strengthen routes that carry useful flow, and you have the beginnings of a transport network that embodies value. In this view, neurons do not appear at the dawn of cognition as a prerequisite; they appear later as one powerful way to internalize and speed up what earlier organisms could do through slower, material loops.

Clark and Chalmers used a notebook because it is a familiar human object, but the slime sheath is, in some ways, an even cleaner example. A notebook is optional; you might forget it. A slime mould cannot forget its ground. The trace is produced automatically, and the reading is integrated into ordinary sensation. The environment is not a peripheral aid; it is an active participant in the control system.

The temptation, at a dinner party, is to say that the slime mould "thinks with the floor". That is catchy, and it is not entirely wrong. Yet the deeper point is that thinking, in a broad biological sense, may simply be what happens when an organism can make its own history matter. Some organisms do that by changing synaptic weights. Some do it by changing hormone levels. *Physarum* does it, in part, by leaving a residue and treating that residue as information. A patch of ground becomes less attractive because it carries the organism's past. Space itself takes on a faint moral quality: *not forbidden, but already tried.*

If this sounds austere, it becomes strangely moving when one watches the behaviour unfold. The organism expands with a kind of earnest curiosity, touching many possibilities, and then, gradually, it lets go. It withdraws from what did not pay. It commits to what did. It does not hoard its entire past inside itself; it lets the world carry some of the burden. The slime trail dries into a thin film, and in that film lies a record of where the organism has had doubts, where it has wasted effort, where it has pressed against a barrier long enough to decide to turn away. The ground keeps the receipt, and the organism keeps moving.

Chapter 7

Algorithms of Life

Nature does not compute with binary logic or silicon chips; it calculates using flow, pressure, and resistance. By translating the slime mold's pulsing veins into mathematical equations, we uncover the "Physarum Solver," a biological algorithm that challenges our traditional understanding of how computation happens in the physical world.

7.1 The Simple Rule of Flow

If you watch *Physarum polycephalum* for long enough, you begin to notice something slightly unnerving: it behaves as though it has a preference. Not a reflex, not a simple twitch away from harm, but a kind of quiet insistence about how to connect places. Put two oat flakes down on a damp surface and the organism will spread, branch, hesitate, and then, over time, commit. Some strands thicken into glossy highways; others fade into thread and vanish. It feels purposeful. Yet inside this yellow sheet there is no brain, no nerve, no command center. The closest thing to a conductor is a pulse.

The pulse is literal. The slime mold is not a colony of cells but a single giant cell, a living bag of cytoplasm threaded through a lacework of tubes. Those tubes rhythmically contract and relax. The contractions are coordinated in waves, like peristalsis in an intestine, sending the internal fluid back and forth through the network. Under the microscope, you can see tiny granules carried along in the streaming flow, reversing direction every minute or so. The organism is constantly pumping itself, not because it enjoys the exercise, but because this circulation moves oxygen, nutrients, chemical signals, and the building materials of its own body. Movement and metabolism are inseparable.

The remarkable thing is what the pulsing flow *does* to the architecture. Each tube has a diameter, and diameter matters. A wide tube offers little resistance: fluid passes easily. A narrow tube is a bottleneck. In a household plumbing system, you do not need intelligence to predict that more water will go through the wider pipe if pressures are similar. The slime mold adds a twist: it remodels its pipes based on how much fluid they carry. Where there is heavy traffic, the tube widens; where there is little, the tube shrinks and may disappear. This is the simple rule of flow.

That rule can be told without mathematics: *use makes strong*. But the organism's "use" is not counted in visits or steps; it is measured in flux, the rate at which cytoplasm moves through a tube. Flux is a physical quantity, and the advantage of physical quantities is that they do not need interpretation. A tube does not have to *know* it is part of a shortest path. It does not have to recognize an oat flake as food in the way an animal recognizes prey. It only has to obey local chemistry and mechanics: stretch a little when fluid shear is high; stabilize and reinforce where contractions repeatedly drive strong flow; let lightly used branches relax and thin.

A concrete historical moment helps fix the idea. Around the turn of the millennium, Toshiyuki Nakagaki and colleagues became famous for a deceptively simple experiment: a slime mold placed in a maze with food at two exits eventually left behind a single thickened route connecting the food sources. The popular retelling makes it sound as though the organism "solved" the maze in one decisive act. The real story, more interesting, is that it began in an almost wasteful way. It spread into corridors broadly, exploring multiple options at once. Only later did it retract from dead ends and redundant loops, concentrating into a minimal connection. What looks like planning is, at the level of the organism, remodeling driven by flow.

Imagine the early maze as a wet network of possible tubes. At first the slime mold fills many corridors with thin veins. The rhythmic pumping then sends cytoplasm through this provisional map. Wherever the two food sites generate stronger metabolic activity—they are sources of attractants and sinks for resources—pressure differences arise across the network. In broad terms, pressure is higher where contractions push and lower where resources are being consumed or where the geometry offers easier escape. Fluid follows pressure gradients. If two different corridors connect the same points, the one offering slightly less resistance will, for a given pressure difference, carry slightly more flow. That "slightly" is enough. More flow triggers widening; widening reduces resistance; reduced resistance increases flow. A path that was only marginally favored becomes strongly favored, not by deliberation but by positive feedback.

The feedback loop is the heart of the matter. It is also a fragile thing, because positive feedback can run away. In a microphone too close to a speaker, a tiny sound becomes a howl. In a slime mold, the loop is tamed by costs. Tissue is not free. Making a tube wider requires material, energy, and opportunity: cytoplasm allocated to one branch is cytoplasm not used elsewhere. Meanwhile, maintaining a wide tube that carries little flow is a waste. The organism therefore behaves as though it is balancing a budget. The widening rule is not a simple ratchet; it competes with constant tendencies to shrink, decay, and reallocate. A living network is always under renovation.

One way researchers captured this is by borrowing language from electrical circuits and fluid dynamics. Think of the network as a set of edges (tubes) connected at junctions. Each tube has a conductance D, a measure of how easily fluid passes through it; equivalently it has a resistance R, with $R \propto 1/D$. Pressure differences play the role of voltage differences. Flux Q plays the role of electrical current. A very simple relationship, akin to Ohm's law, links them:

$$Q_{ij} = \frac{D_{ij}}{L_{ij}}(p_i - p_j),$$

where Q_{ij} is the flux through the tube between nodes i and j, p_i and p_j are pressures at the nodes, and L_{ij} is the tube's length (or more generally the cost of traversing that segment). This is not meant as a perfect micro-scale description of a pulsing, viscoelastic gel. It is a deliberately stripped-down caricature. The point is that, given a set of conductances, the physics of flow determines the fluxes.

The biological innovation enters when conductances are allowed to change. A common modeling choice, inspired by experiments and formalized elegantly by Atsushi Tero and colleagues in the mid-2000s, is a differential equation of the form

$$\frac{dD_{ij}}{dt} = f(|Q_{ij}|) - \mu D_{ij}.$$

Here $f(|Q_{ij}|)$ is an increasing function: larger flux reinforces the tube. The term μD_{ij} represents decay: in the absence of reinforcement, tubes relax and shrink. Different choices of f tune the behavior. A popular option is a power law, $f(|Q|) = |Q|^\gamma$, where γ controls how strongly reinforcement favors already-busy tubes. If γ is small, even modest flows keep multiple routes alive; the network stays redundant. If γ is large, the rich get richer: one route dominates and others wither. These are not merely mathematical flavors; they correspond to visibly different morphologies in real slime molds, which can form either sparse tree-like networks or richly looped webs depending on conditions.

The seductive part is how much global structure falls out of this local equation. Pressures and flows are global in the sense that they depend on the entire network, but no tube needs to compute the whole. Each tube only "feels" its own flux and its own mechanical and chemical state. Yet the coupled system pushes toward networks that look optimized: short connections when efficiency dominates, more loops when robustness matters, thicker main veins where traffic is heavy, thin exploratory filaments at the edges.

The visual intuition is worth making explicit, because it is easy to miss why this is different from a mere river eroding a channel. Erosion is also flow shaping structure. The difference is that the slime mold's channels are alive, actively pumped, and reversible on timescales of minutes to hours. It can build and erase plumbing repeatedly as circumstances change. It is closer to a city that widens roads where there is congestion and narrows them where traffic disappears, doing this continuously and without central planning, using only local measurements.

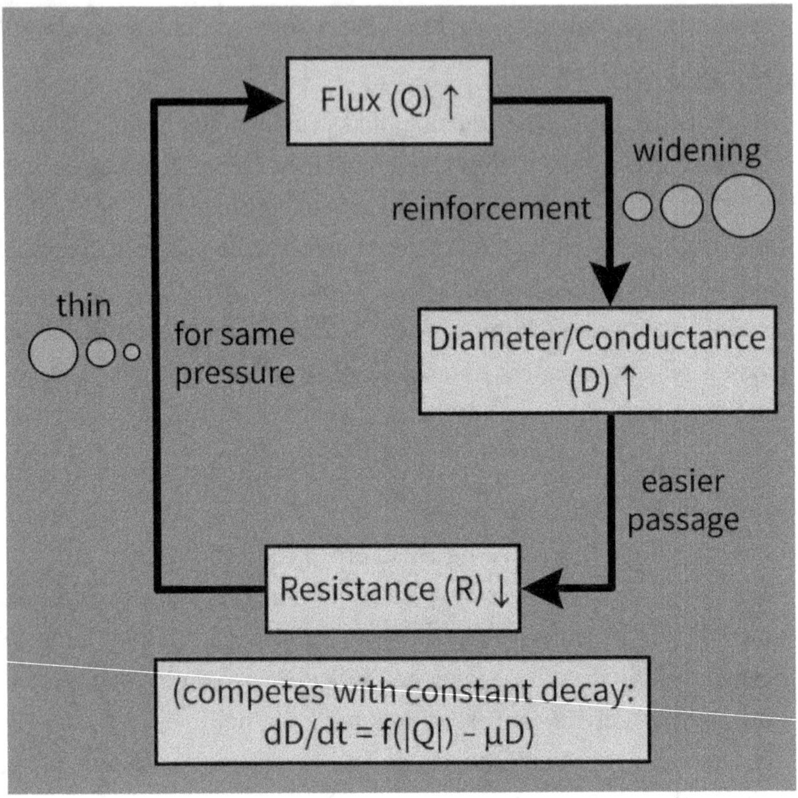

A simple loop like this is sometimes mistaken for a mere trick of amplification, but it has a deeper consequence: it turns the network into a physical memory. Where flow has been, structure remains, at least for a while. The slime mold does not store a list of visited corridors; it stores widened veins. This kind of memory is not symbolic. It is closer to a footpath in grass. Walk the same route repeatedly and you do not need to remember it consciously; the landscape remembers for you.

That memory becomes computation when combined with the organism's repeated sampling of alternatives. The initial web provides many potential routes. The pulsing circulation tries them all in parallel. Slight differences in resistance produce slight differences in flux. The reinforcement law magnifies those differences. Decay deletes the losers. After enough time, the remaining structure encodes a solution to a problem posed by the environment: connect these food sites; do so with minimal cost given the geometry and constraints.

A small, tangible example makes the cost-efficiency tradeoff feel less abstract. Consider two oat flakes placed on opposite sides of a petri dish, with a small obstacle in between: perhaps a dry patch or a piece of plastic. The slime mold spreads around the obstacle, forming two arcs. One arc is slightly longer; perhaps the dish edge forces it to curve. At first, both arcs contain tubes. The organism pumps cytoplasm between the food sites, and both arcs carry some flow. But the shorter arc has lower effective length L, so for similar pressures it carries more flux. Its conductance grows. As it grows, even more flux is diverted into it, like a shortcut that becomes more attractive because it is now a paved road rather than a dirt track. The longer arc becomes underused; its conductance decays; eventually it is so thin that flow there becomes negligible. The organism has not only found the shorter connection but invested its material into it.

This can look almost too neat, as though the world always presents a single best route. Real environments are messier. Food sources are not equally valuable. Conditions shift: light, humidity, toxins, competitors. A purely shortest-path obsession would be dangerous. A thin, single tube might be efficient but brittle: one cut and the organism is split. Here the parameter choices in the reinforcement function matter, and so does the biology. *Physarum* often maintains loops, especially in richer environments where risk is worth hedging. In the language of networks, loops provide fault tolerance. If one tube is damaged, flow can reroute. The same reinforcement rule can yield either tree-like frugality or web-like resilience depending on how strongly flux is allowed to dominate decay and how the organism's physiology responds to stress.

Even the back-and-forth nature of streaming, which sounds at first like indecision, helps. Because the flow reverses, the organism does not simply push resources one way and starve a branch permanently. It repeatedly tests conductances under alternating conditions. A tube that is occasionally useful can be kept alive if it regularly carries some flux; a tube that is consistently useless, even if it once had a burst of activity, will fade. The system is not optimizing in one shot; it is continually re-optimizing.

This is where the "intelligence" claim should be treated carefully. Intelligence, in everyday speech, implies a mind contemplating options. What the slime mold shows is that impressive problem-solving can arise from an arrangement of matter with the right feedback. The organism senses and acts through the same medium. Chemical gradients influence contraction patterns; contraction patterns drive flows; flows redistribute chemicals and mechanically stress tube walls; stressed walls remodel; remodeled walls change the next pattern of flow. A loop of causation replaces a loop of reasoning.

There is also something quietly human about the way this works, which may explain why it fascinates beyond biology. The rule resembles what engineers call a *reinforcement* or *adaptive* algorithm: increase capacity where demand is high, reduce it where demand is low. Anyone who has cursed at a traffic jam and wondered why the road was not wider has implicitly appealed to the same principle. Yet in human systems, data must be gathered, budgets negotiated, plans approved, contractors hired. The slime mold's bureaucracy is chemistry. Its planning office is physics.

That does not make it infallible. A positive feedback loop can lock in early accidents. If the organism happens to reinforce a suboptimal route first, it may take time to undo that choice. A tube that becomes wide can dominate flow even if a slightly better alternative exists elsewhere, simply because it has already lowered its resistance. The decay term is the mechanism for forgetting, and forgetting takes time. This, too, feels familiar: the first road built between two towns often shapes development for centuries, not because it was the best possible road, but because once it exists it attracts traffic, commerce, and further construction.

The power of the simple rule of flow lies in its minimalism. It does not require the organism to represent the maze, or to rank options, or to compute distances. It requires only that matter rearranges itself in response to what passes through it, with a steady pressure of decay preventing the system from ossifying into yesterday's solution. The result is a network that looks as though it is thinking, because it embodies the environment's constraints in its own shape. In a slime mold, solving a problem is not separate from becoming a particular form; the answer is written in tube thickness, traced in living gel, and sustained only so long as the pulse keeps moving.

7.2 The Physarum Solver

A certain kind of scientific romance begins when an organism behaves like an engineer. After watching a slime mold reinforce some veins and abandon others, it is hard not to ask a very human question: can that behavior be written down as an algorithm?

The question matters because it runs in both directions. On one side, a good mathematical model explains how something that looks like choice can emerge from physics and local feedback. On the other, the same model can become a practical tool: a new way to route traffic through a communication network, to plan a supply chain, or to design wiring on a chip. The organism lends intuition; mathematics lends portability.

The phrase *Physarum solver* names a family of models that translate the slime mold's adaptive plumbing into equations on a graph. A *graph*, in this sense, is not a plot with axes. It is a network: a collection of nodes (junctions) connected by edges (links). Nodes can represent food sources, intersections, routers, or cities. Edges represent corridors in a maze, roads between towns, or fiber-optic cables. Each edge has a *length* or *cost*—the distance, time, energy, or money required to traverse it. The goal of many classic problems in computer science and operations research is to find a cheap way to connect things in such a graph.

One of the simplest is the *shortest path* problem: given a start node s and a target node t, find the least-cost route connecting them. It is old enough to feel almost like common sense, and it sits inside countless technologies: GPS navigation, packet routing on the internet, and the scheduling of deliveries. The shortest path problem is also, from the standpoint of computation, well behaved: algorithms like Dijkstra's method can solve it efficiently even for very large networks.

Harder problems begin when there are more than two terminals. Suppose there are several important nodes that must be connected: multiple warehouses, multiple servers, multiple cities. One wants a network that joins them at minimal total cost. If you insist that the network can only use the given nodes, the task is called the *minimum spanning tree*: connect all nodes with minimal total edge length while avoiding cycles. This too can be solved efficiently.

But real networks often do something subtler. They create junctions where it helps—branch points that were not pre-designated as "important" locations. You can see this in a road network that adds an interchange, or in plumbing that introduces a manifold, or in wiring that uses intermediate points to reduce total copper. When a network is allowed to include extra junctions, called *Steiner points*, the problem becomes the *Steiner tree* problem. Steiner trees are notorious:

they are computationally difficult in the formal sense used by computer scientists (NP-hard). "NP-hard" does not mean impossible, but it warns that no efficient method is known for guaranteeing the best answer in all cases. For large networks, one typically uses heuristics: methods that often work well but do not promise perfection.

Slime molds enter this story because their behavior looks like a heuristic for precisely these network design tasks. Put oat flakes down at several points on a dish and *Physarum* tends to build a network of tubes that links them. The resulting network is neither a single minimal tree nor a maximally redundant web. It often strikes a compromise: relatively short total length, but with alternative routes that provide resilience if part of the network is damaged. This compromise shows up vividly in the much-discussed experiment by Atsushi Tero and colleagues (2010), who compared *Physarum*'s networks connecting food sources laid out like the Tokyo rail system to the actual rail network. The resemblance was not exact, and it should not be overstated, but it was compelling enough to make a point that is easy to feel: a living network can make trade-offs that engineers care about.

Turning that biological intuition into a solver begins with the same basic ingredients as the simple feedback rule, but now phrased on an abstract graph. Each edge e has a length (cost) $L_e > 0$ and a time-dependent conductance $D_e(t) \geq 0$ describing how "thick" or "capable" that edge is. There is a pressure value p_v at each node v. If fluid is injected at a source node s and removed at a sink node t, then conservation of mass requires that flow neither appears nor disappears at intermediate nodes. In the simplest steady formulation, one unit of flow enters at s, one unit leaves at t, and everything in between balances.

Given conductances, the flows are determined by a network version of Ohm's law:

$$Q_e = \frac{D_e}{L_e} \Delta p_e,$$

where Δp_e is the pressure difference across edge e (with a sign depending on direction). The division by L_e is a convenient way to make long edges intrinsically less attractive: for the same conductance and pressure drop, a longer edge carries less flow. If one prefers a more physical picture, D_e / L_e acts like an effective permeability of that connection.

The "Physarum" ingredient is that D_e adapts to $|Q_e|$. A standard choice is

$$\frac{dD_e}{dt} = |Q_e| - \mu D_e,$$

with $\mu > 0$ setting the timescale of decay. This is a stripped model of reinforcement and atrophy: heavy use thickens the edge; unused capacity fades. One can replace $|Q_e|$ with a more general increasing function $f(|Q_e|)$ to tune how sharply the system commits to a dominant route. The exact functional form is less important than the coupled logic: flows depend on conductances, and conductances evolve in response to flows.

A remarkable fact, made precise by Vincenzo Bonifaci, Kurt Mehlhorn, and collaborators (2011) and sharpened in later proofs, is that this continuous-time dynamics does not merely look like it is searching for a good path. Under standard assumptions, it converges to the shortest path between s and t in a weighted graph. That statement is both stronger and stranger than it first appears. It says that a system that never explicitly compares complete routes, never enumerates paths, and never stores distances can nevertheless settle into a state where essentially all conductance concentrates on the minimum-cost path. The "solution" is not printed out as a sequence of nodes; it exists as a pattern of thickness.

One way to make this believable is to picture what the dynamics is doing to the network over time. At the beginning, suppose every edge has some small conductance, as if the slime mold has laid down a thin exploratory film everywhere. With a unit of flow forced from s to t, the network distributes that flow across multiple routes. Any edge that lies on many plausible routes carries some of the load; edges on dead ends carry almost none. Now an asymmetry begins: an edge on a shorter overall route tends to carry slightly more flow, because the total resistance of that route is lower, and parallel resistors split current inversely to resistance. The adaptation rule amplifies that asymmetry by increasing conductance where flow is high. Increased conductance further reduces resistance on that route, diverting even more of the unit flow to it. Competing routes lose flow and shrink. The system behaves like a tournament in which "fitness" is measured by carried flux, and flux is itself shaped by the current state of the competition.

There is also a more mathematical way to gain intuition. The total "effort" required to send one unit of flow from s to t through a resistive network is related to an energy:

$$E = \sum_{e} \frac{L_e}{D_e} Q_e^2.$$

Edges with high resistance (low D_e) make energy expensive; pushing a lot of flow through them costs. The adaptive rule tends to allocate conductance to edges that carry flow, which can be interpreted as the network lowering the energetic cost of moving material between the terminals. In certain formulations, the dynamics can be connected to an optimization process: it resembles a kind of steepest descent on a landscape defined by the network's constraints. Damian Straszak and Nisheeth Vishnoi (2015) pushed this viewpoint further, linking Physarum-like dynamics to linear programming, suggesting that what began as a biological metaphor may be one instance of a broader class of natural optimization processes.

Shortest path is only the beginning, though it is the cleanest case because there are only two terminals and the physics can be anchored to a single unit of flow. To move toward Steiner-like problems, one replaces the single source and sink with multiple sources and sinks, or with multiple demands: several pairs of nodes may need connection, or one may need to distribute flow from one source to many destinations. Another approach is to run multiple source—sink problems and let the conductances adapt under the combined traffic. Edges that are useful for many demands tend to be reinforced because they carry high aggregate flow; edges that only help rarely decay. The resulting network can approximate a Steiner tree, because Steiner trees are precisely about building shared trunk lines that serve multiple terminals efficiently.

A concrete example makes this less abstract. Imagine three data centers that must exchange information: nodes A, B, and C. The naive approach is to build three separate shortest paths for each pair: A—B, B—C, and A—C. That can be wasteful, duplicating infrastructure. A more economical network might build a shared backbone: a path from A to some junction J and then branches to B and C. The junction J might not be a data center; it is a Steiner point, justified purely by savings. In a Physarum-style adaptation, if traffic is substantial between all pairs, then edges that lie on a shared backbone carry more total flow because they serve multiple exchanges. Those edges get reinforced, effectively "discovering" that a shared trunk is valuable. Where the demands are uneven, the solver does something else: it thickens routes that serve the heavy demands and allows light demands to be served by thinner, more peripheral links, again reflecting a budget.

This is close to how the living slime mold seems to behave when presented with multiple food sources. The organism does not simply connect them with a minimal tree and stop. It keeps pumping. That pumping means the network is always being tested under load. If a particular branch is essential for distributing nutrients, it will carry a lot of flow and thus remain thick. If a shortcut is rarely used, it may thin but not necessarily vanish, especially if the reinforcement function is gentle enough that moderate flows can maintain secondary routes.

The most striking feature of the Physarum solver, for engineers, is that it is naturally *distributed.* A classical shortest path algorithm such as Dijkstra's method assumes a kind of centralized bookkeeping: a table of tentative distances is updated by selecting the smallest currently known distance, expanding outward in a disciplined order. This is very efficient on a computer because a computer can rapidly store and compare numbers. But in a decentralized system—a swarm of robots, a sensor network scattered across a forest, a set of devices with limited communication—that style of global bookkeeping is awkward. One can distribute it, but the distribution is an additional design challenge.

In a Physarum-style solver, distribution comes for free because the "computation" is embedded in local adaptation. Each edge updates its conductance using only information about its own flux, and flux is a consequence of physical constraints propagated through the network. Pressure is not a global instruction; it is an emergent field, like voltage in a circuit, determined by conservation laws. This resemblance is more than poetic. Engineers already solve many optimization and routing tasks by building analogies to resistive networks. What the Physarum solver adds is an adaptive element that turns a static analog into a self-improving one.

There is an honest limitation here, and it is worth keeping in view: the Physarum solver is not a magical shortcut through computational complexity. For NP-hard problems like the Steiner tree, no one expects a simple biological metaphor to guarantee optimal solutions for all cases. What it offers is a principled heuristic with a biological pedigree, and sometimes that is exactly what is needed. Real-world network design rarely demands a provably optimal solution; it demands a good solution that can be found reliably under changing constraints. When links fail, when costs shift, when new terminals appear, a network that can adapt continuously may be more valuable than one perfect plan computed once.

The same adaptive character also explains why the model became a bridge between biology and theoretical computer science. When a group like Bonifaci and Mehlhorn shows convergence to shortest paths under a dynamical system inspired by slime mold, it is not primarily about helping GPS apps (which already work well). It is about demonstrating that a continuous, local, physically plausible process can be mapped onto a discrete computational guarantee. That mapping changes one's sense of what an algorithm can be. An algorithm need not be a sequence of symbolic instructions executed step-by-step. It can be a set of coupled differential equations that relax toward a solution, using the world's own laws as the computing substrate.

A final anecdote brings the abstraction back down to the bench. In many laboratories, the first time someone tries a slime mold maze experiment, they are disappointed. The organism does not "choose" immediately. It smears itself into multiple corridors, sometimes even appearing to ignore the food. Then, hours later, the network begins to look decisive. The disappointment is a misunderstanding of the time scale. The solver is not a snap judgment; it is an iterative physical process. The organism is continually sending test currents through its own body, continually updating conductances, continually paying maintenance costs. The answer arrives not as a single act but as a shape that becomes increasingly difficult to unmake.

When the conductances have settled, the network has, in effect, performed a piece of arithmetic with geometry: it has taken a collection of possible connections and compressed them into a sparse set that can carry the required flow at low cost. That arithmetic is written in thickness, sustained by pulse, and constrained by the same trade-offs that shape human infrastructure: efficiency bought with material, resilience bought with redundancy, and a constant pressure toward parsimony whenever a route falls quiet.

7.3 Computing with Matter

A computer, in the ordinary imagination, is a box that runs instructions. Software is intangible, hardware is solid, and intelligence lives in the separation between the two: a pattern of symbols manipulating other symbols. Slime molds tempt a different picture. When *Physarum* finds a good route between food sources, no list of candidate paths is stored anywhere. There is no internal "map" that can be read out like a screen. The answer is a shape: a thickened tube that persists because it is continually used, and that vanishes when it is not. What looks like problem-solving is inseparable from the organism's material.

That idea has a name that is both helpful and misleading: *morphological computation.* "Morphological" comes from morphology, the study of form. The claim is that a body can do computational work simply by being what it is, placed in the right environment. Springs, dampers, friction, elasticity, and geometry can take over tasks that would otherwise need explicit control and calculation. Instead of a central processor telling every part what to do, the parts constrain one another so tightly that desired behavior emerges naturally.

Engineers have been using this principle for a long time, often without the label. A passive dynamic walker, for example, is a pair of mechanical legs that can walk down a gentle slope with no motors and no computer, purely because its joints and mass distribution cause it to swing in a stable way. The "control" is built into the mechanics. The walker does not compute in the sense of performing arithmetic, but it does something a control algorithm is typically hired to do: it stabilizes a gait. The environment and the morphology conspire to produce the result.

Slime molds push the same idea into the realm of networks and optimization. Their body is not a rigid skeleton but a reconfigurable graph of tubes. Its physics carries information. Pressures propagate. Flows sum and cancel. Thickening and thinning turn transient currents into persistent structure. Instead of computing a shortest path and then building it, the organism builds many possibilities, tests them under load, and lets the test reshape the builder. If one insists on the hardware—software division, *Physarum* looks like a paradox: the "program" is an oat flake and a humidity gradient.

This is where unconventional computing, a field associated particularly with Andrew Adamatzky and collaborators, enters with an almost mischievous ambition. Why restrict computation to silicon? If a slime mold's protoplasmic network can embody a solution to a routing problem, perhaps it can be used as a physical substrate for computation itself: not merely as a metaphor, but as an actual device whose outputs can be measured and used. In this view, the slime mold is not a mascot for distributed intelligence. It is a component.

The first move in this direction is surprisingly simple. The organism is sensitive to chemical cues (food, salts, toxins), to light, to moisture, and to temperature. Arrange attractants and repellents in space and the slime mold's growth pattern changes. That pattern is readable: one can photograph it, measure its thicknesses, track the timing of its oscillations, or record changes in electrical potential across the body. If different inputs produce reliably different outputs, then, in a broad sense, the organism can be said to *compute* a function from inputs to outputs.

A concrete example has an almost Victorian charm. Place the slime mold in a petri dish with several oat flakes arranged as the corners of a polygon. The organism will tend to connect them with a network that approximates something like a minimal connection. Shift one oat flake slightly and the network reorganizes. The output is not a number but a geometry. Yet that geometry can be useful, and it can be fed back into human design. A lab might use image processing to extract a graph from the slime mold's tubes and then propose that graph as a candidate design for wiring or for a transportation network. The organism becomes, in effect, an analog suggestion engine.

More daring demonstrations treat *Physarum* as a logic element. Logic gates are the building blocks of digital computation: AND, OR, NOT. In silicon, they are realized through transistors. In a slime mold, one can try to realize them through growth and avoidance. If two inputs correspond to placing attractants at two locations, the organism may or may not connect them depending on the configuration, producing an output connection elsewhere. These devices are slow, fragile, and more whimsical than practical. Their value is conceptual. They remind us that "computation" is not tied to electrons and clock cycles. It is tied to reliable transformations of state.

The most persuasive claim for computing with matter, though, does not rest on building a slime-mold laptop. It rests on seeing that physical processes can solve problems in the same family as those solved by algorithms, sometimes with an elegance that is hard to match symbolically. When the adaptive network dynamics described earlier converges to a shortest path, it is already a mathematical version of this idea: a computation realized as relaxation in a physical-like system. In the real organism, the relaxation is implemented by streaming cytoplasm and remodeling tube walls. The "variables" are diameters and pressures. The "constraints" are conservation laws and material costs. The "output" is a stable transport network.

Calling this computation is tempting, but it also risks turning a useful insight into a slogan. Critics of morphological computation have argued, quite reasonably, that if every physical process is called computation, the word loses meaning. A rock rolling downhill "finds" a low point; a soap film spanning a wire frame "solves" a minimal surface problem; water flowing through a landscape "computes" a drainage basin. These are beautiful analogies, but do they really count as computing? Or are they simply physics doing what physics does?

A way through the debate is to pay attention to what is being claimed. There is a modest claim and a bold claim.

The modest claim is that morphology can *simplify control.* Instead of a controller calculating every correction, the body's dynamics provides stability and responsiveness by itself. That is clearly true in many systems, from animal locomotion to compliant robotic grippers. The body does not compute an equation; it embodies one.

The bold claim is that morphology can *implement general computation*, comparable in scope to what a Turing machine can do. A Turing machine is a theoretical abstraction that captures the idea of a programmable computer: a head that reads and writes symbols on a tape according to rules. Its power lies in universality: with enough tape and time, it can compute anything that is computable in principle.

Slime molds, at least as currently understood and used, sit closer to the modest claim. They are excellent at certain kinds of spatial, analog tasks: routing, pattern formation, distributed sensing. They are not obviously suited to arbitrary symbolic manipulation. Their strengths are also their constraints: they live in continuous space, they are noisy, their "operations" are slow, and their internal states are coupled to metabolism. If one asks them to behave like a digital circuit, one is forcing them into a costume that does not fit.

And yet the bold claim keeps reappearing, partly because it touches a deep philosophical nerve. If a computation can be realized by a physical system relaxing to equilibrium, then the boundary between computation and physics blurs. Many optimization problems can be framed as minimizing some cost function, and many physical systems can be framed as minimizing energy. Soap films do not know calculus, but they are very good at reducing surface area. A slime mold does not know graph theory, but it is very good at reducing transport cost while maintaining flow. One begins to wonder whether the universe is full of "solvers" that simply happen to be made of different substances.

A historical anecdote helps here, because analog computation has been tried before, at scale, with far less romance than slime molds. In the mid-twentieth century, before digital computers dominated, engineers used analog machines that represented variables as voltages and solved differential equations by wiring together components. These devices could simulate aircraft dynamics or electrical circuits in real time. They were powerful, but they were also finicky: noise, drift, and limited precision were constant problems. Digital computers won not because analog machines were conceptually wrong, but because digital machines were easier to copy, scale, and program reliably.

Slime-mold computing resembles analog computing in this older sense. It uses continuous variables (tube thickness, flow, chemical concentration) and relaxes toward a useful state. It can be robust in a biological way—repairing itself, tolerating local damage—but it is also temperamental, sensitive to humidity and contamination, and slow by electronic standards. Its value is therefore not in competing with silicon, but in widening the repertoire of what we count as information processing and in inspiring new kinds of algorithms that borrow the virtues of matter: parallelism, adaptability, and graceful degradation.

The phrase "offloading computation to the body" captures the engineering appeal. Suppose one wants a swarm of tiny robots to search a collapsed building for survivors. If each robot must build a map, compute routes, and communicate detailed plans, the burden of computation and coordination is heavy. But if the robots can lay down and sense a chemical trail, like ants, and if their motion rules cause trails to reinforce where traffic is useful and fade where it is not, then the environment becomes part of the memory. Coordination is achieved not by centralized planning but by stigmergy: indirect communication through traces in the world. The slime mold does something similar internally: its veins are both paths and records of use.

There is a subtle moral here about intelligence itself. In the familiar brain-centered view, intelligence is internal. The world presents a problem; the brain computes a solution; the body acts it out. In a morphological view, intelligence is distributed across brain, body, and environment. The "solution" may be partly stored in muscle elasticity, partly in the layout of a nest, partly in a chemical gradient that persists. What *Physarum* dramatizes is that even without neurons, a system can accumulate information in the structure of its own material and use that structure to guide future behavior.

One can make this almost embarrassingly personal. Consider how people arrange their kitchens. The items used most often end up in the easiest-to-reach places; the rarely used gadgets migrate to back shelves. No one runs an optimization program to compute the arrangement. The arrangement emerges from repeated use and convenience, reinforced by habit. The kitchen becomes a kind of external memory of daily life. A slime mold's network is a kitchen that rearranges itself continuously, with flow acting as the hand that reaches for what matters.

Seen this way, the excitement around slime molds is not only about a clever trick with mazes. It is about a shift in what counts as a computational resource. In a digital computer, matter is carefully disciplined. Electrons are channeled through transistors in ways that suppress most of the physics, leaving only crisp binary states. In computing with matter, one does the opposite: one recruits as much physics as possible. Viscosity, diffusion, elasticity, noise, and even decay become part of the algorithm. The price is loss of precision; the reward is natural adaptation.

There is still a question that refuses to go away: where, exactly, is the boundary between "the body helps" and "the body computes"? A useful practical answer is to look for *programmability* and *readout*. A process becomes computational in an engineer's sense when it can be steered by inputs into reliably different outcomes, and when those outcomes can be interpreted as answers to a class of questions. A slime mold can be programmed crudely by placing attractants, repellents, barriers, and nutrients. Its readout is the geometry and thickness of its network, the timing of its oscillations, and measurable electrical signals. These are not universal capabilities, but they are enough to make the organism a genuine information-processing system rather than a mere metaphor.

Perhaps the most honest description is that *Physarum* offers a demonstration of a middle ground: computation that is not symbolic but not mindless either; not designed but not random; distributed yet coherent. It shows how a body can be both the machine and the message. When the organism thickens a tube, it is simultaneously building infrastructure, moving resources, sensing the environment, and recording a history of use. The same act serves multiple purposes because the system is not partitioned into modules. Life rarely is.

On a petri dish, the result can look like a delicate map drawn in living ink. The map is not a representation of the world; it is a negotiated truce between the world's constraints and the organism's costs, refreshed with every pulse of cytoplasm. In that truce lies a different vision of intelligence: not as a hidden calculation, but as a form that remains poised to change.

Chapter 8

Engineering the Future

Biology often solves complex topology problems long before humans draft the blueprints, offering a template for efficiency that defies conventional engineering. This chapter moves from the petri dish to the city street, exploring how the slime mold's rules of engagement are reshaping resilient infrastructure and inspiring a new generation of soft, decentralized robots.

8.1 The Slime Mold City Planner

A railway map has an uncanny way of looking inevitable. Lines radiate from a city center, branch, thicken, and loop as if geography itself demanded that particular arrangement. Yet every link on that map is the end of arguments: where to spend scarce money, which neighborhoods to connect first, what risks to accept if a bridge fails or a tunnel floods. The mathematical version of this problem is older than railways. Given a set of important places ("nodes"), how should one connect them with paths ("edges") so that travel is fast and construction is cheap?

If you ask classical graph theory to be ruthlessly efficient, it tends to give a brittle answer. In its purest form the problem becomes a hunt for a minimal network: connect all the nodes with the least total length of wire, pipe, track, or road. One famous ideal is the *Steiner tree*, named after the nineteenth-century geometer Jakob Steiner. The Steiner tree is like a thrift-obsessed engineer: it is willing to introduce extra junctions (Steiner points) if that lets the overall network shrink, but it refuses to build any redundant loops. Every meter of material must "earn its keep" by being essential to connectivity.

That refusal is precisely the problem. A tree, by definition, has no cycles. Remove a single edge, and something becomes disconnected. In the tidy world of blackboard geometry, disconnection is a minor inconvenience; in the world of storms, strikes, accidents, and cascading failures, it is the difference between a delay and a disaster. A city can tolerate a detour, even an ugly one. It cannot tolerate a severed lifeline.

Engineers have long known this in their bones. If you have ever wondered why a laptop charger cable fails at its thin, stressed point rather than in the thick middle, you have met a simple rule: systems break where they are most constrained. Infrastructure behaves similarly. A minimal network concentrates flow into a small number of critical segments, which become hot spots for overload and single points of failure. Human planners respond by adding redundancy: alternate lines, ring roads, bypass valves, spare capacity. The trouble is that redundancy is expensive, and the math quickly becomes unfriendly. As soon as you move beyond "shortest total length" to include resilience, you are no longer solving one neat optimization problem; you are negotiating among several goals that tug in different directions.

This is where a slime mold enters the story with disarming calm.

Physarum polycephalum is not a mold in the everyday sense and not an animal either, but a giant single cell that can spread across a petri dish like a living sheet. It builds networks of veins: thicker tubes that carry more flow, thinner ones that wither if they are not useful. The cell has no brain and no master plan. Its "decisions" are embodied in the very material of its body, guided by local cues and the physics of transport. When food appears at different spots, Physarum grows toward it, connects the patches, and then remodels the connections as conditions change.

Around 2010, Atsushi Tero and colleagues set up a now-famous demonstration that made the point vivid. They placed oat flakes (a favored food) on a plate in the approximate pattern of cities around Tokyo, using the capital as the central hub. The slime mold was introduced at the Tokyo position. Over time, the organism explored and then consolidated into a network of tubes connecting the "cities." The striking part was not that it connected them at all, but that the final network resembled key features of the real Tokyo-area railway system: a backbone of major routes, plus additional links that created alternative paths.

The story is often told as a charming novelty: the slime mold "re-designed Tokyo." The deeper lesson is more technical and more useful. Physarum is implicitly balancing multiple objectives. It wants to deliver nutrients efficiently (short paths and low transport cost), but it also wants to keep the system robust (so that damage or blockage does not starve part of the cell). Those goals are not separate modules; they are fused in the organism's adaptation rule. Tubes that carry more flow are reinforced. Tubes that carry little flow are not. If one route becomes slightly better, it gains flow and strengthens, but alternative routes are not instantly abandoned if they still carry enough flow to justify their upkeep. The resulting network is neither a pure tree nor an indiscriminate mesh. It looks like what a cautious planner might draw after a few painful lessons: *good enough* efficiency with redundancy in the places where redundancy matters.

A planner might object: "Of course a living organism likes redundancy. It cannot call a repair crew." Exactly. The organism is solving a harsher problem than most civil engineers face. A rail operator can dispatch workers after a storm. A slime mold cannot. It must survive by continual, local self-repair, which means it has evolved rules that treat resilience as a default, not an add-on.

The contrast with classical optimization becomes sharp if you imagine two maps on the same set of cities. One is a Steiner tree, drawn with thin, straight segments and a few elegant junctions. It is beautiful, parsimonious, and doomed the first time an edge is cut. The other is a Physarum-like network: the main routes are thick, because they carry the bulk of the flow, but there are loops and cross-links, some thin and tentative, some thick enough to serve as real alternatives. The second map contains waste, if waste means "material that is not strictly necessary in perfect conditions." Yet that "waste" is what allows the system to keep functioning when conditions are not perfect, which is to say, almost always.

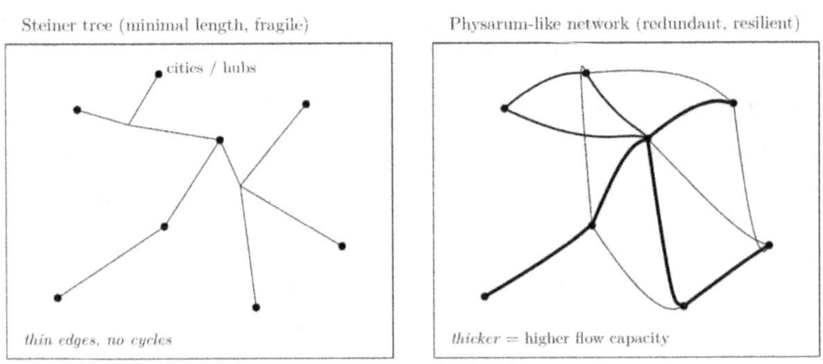

The virtue of the Physarum approach is not mystical intelligence; it is a practical algorithm hiding in plain sight. The organism behaves like an adaptive network designer because its survival depends on moving resources through space while paying a metabolic "rent" to maintain its tubes. Each tube has a cost. Each tube also has a benefit: it can carry flow. If flow increases, reinforcement makes the tube more conductive; if flow decreases, the tube shrinks. That is a feedback loop you could write down as equations, and Tero and collaborators did exactly that in 2010, translating the biological intuition into a mathematical model of adaptive conductance. The model is simple enough to simulate and flexible enough to tune. Turn one parameter and the network becomes more tree-like, emphasizing low cost. Turn it the other way and it becomes more richly meshed, emphasizing robustness.

This tunability matters because real infrastructure planning is never a single-number contest. A water utility cares about energy use (pumping costs), construction and maintenance, vulnerability to pipe breaks, and service equity. A logistics company cares about shipping time, fuel costs, inventory buffers, and the ability to reroute when ports close or borders tighten. A power grid operator cares about line losses, congestion, stability margins, and the nightmare scenario of cascading failure.

Traditional graph theory offers clean answers when the question is clean. Minimum spanning trees, shortest paths, Steiner trees: these are beautiful tools, and they are still used. But they become uncomfortable when the objective is multi-headed. The moment you say, "Minimize length but also minimize the worst-case damage from any single failure," you are no longer just finding a shortest structure; you are exploring a landscape of compromises. The blunt way to do that is to choose weights: decide that a minute of travel time equals some fraction of a dollar of construction, add everything up, and optimize the sum. The trouble is that the weights are political as much as technical. They embed values: whose delays count, what risks are acceptable, which neighborhoods can be left with a single route.

Physarum, in contrast, does not optimize a single scalar objective. It behaves more like a system that must remain viable under uncertainty. Instead of declaring a fixed trade-off in advance, it keeps options alive until the evidence of use (flow) makes a decision unavoidable. That is a very different style of computation. It is not "find the best network"; it is "keep a network that works, then adapt as demands and constraints reveal themselves."

One way to see the distinction is to think about how failure is represented. In a Steiner tree, a failure is catastrophic: an edge is removed and the graph disconnects, full stop. In a Physarum-like network, failure is just another perturbation. Cut a tube, and flow redistributes through remaining pathways. That redistributed flow then reinforces the alternatives, thickening what used to be a secondary route. The system does not merely *use* redundancy; it *allocates* redundancy dynamically, creating a memory of where stress occurred. Engineers often try to add this after the fact through monitoring and reinforcement programs. The slime mold bakes it into the network's material rules.

A concrete historical analogy comes from the early days of London transport. The first underground lines in the nineteenth century were not designed as a coherent, redundant system. They were built by competing companies, focused on getting passengers from one profitable corridor to another. The map that emerged was fragmented and awkward. Over time, as ridership patterns and failures exposed weaknesses, links were added and interchanges improved. Redundancy came slowly, often after disruptions made the cost of fragility visible. What Physarum offers is a caricature of that long, expensive learning process compressed into a growth-and-prune dynamic: explore broadly, then consolidate, but do not consolidate so far that the system cannot flex.

The Tokyo rail comparison was compelling precisely because Tokyo is not a tidy theoretical city. It is a dense, historically layered metropolis with complex demand patterns. The resemblance suggested that there is something structurally sensible about networks that are not minimal, yet not fully meshed. They are shaped by traffic, constrained by cost, and tested by perturbations. Physarum does not know Tokyo, but it does know what it means to keep a sprawling body fed when the world pushes back.

The temptation, especially in popular retellings, is to declare a victory of biology over mathematics. That is too simple, and it misses the engineering value. Physarum is not a magic oracle that outputs the correct railway plan. It is a generator of *candidate networks* that sit on a useful frontier: reasonably efficient, reasonably robust, adaptable. That frontier is what planners actually need. When budgets tighten, they can slide toward efficiency. When climate risks rise, they can slide toward redundancy. Tero's model shows that by adjusting parameters you can traverse a family of designs rather than committing to a single brittle optimum.

This is also why slime-mold-inspired algorithms are attractive for infrastructures that must cope with uncertainty by design. Supply chains are an obvious example. A just-in-time chain is the logistical equivalent of a Steiner tree: lean, cost-efficient, and liable to snap when one port closes or one factory goes offline. The recent history of global shipping has made that fragility painfully concrete. The more resilient alternative looks less like a single perfect route and more like a network with slack: multiple suppliers, alternate transport modes, warehouses that seem "wasteful" until they are suddenly essential. In that setting, an algorithm that naturally produces redundant loops and adapts flow to changing conditions is not a curiosity; it is a prompt to rethink what counts as optimal.

Power grids pose an even sharper version of the same issue. Electricity flows do not politely follow a chosen route the way cars do; they distribute according to physical laws, and local failures can cascade. Grid operators therefore value meshed topologies and controllable redundancies. Yet building and maintaining extra lines is expensive, and the best topology depends on demand patterns that shift with weather, economic activity, and the electrification of transport. A Physarum-like approach offers an intuition: allow the network to "feel" load by representing it as flow, then reinforce pathways that carry stress while letting unused capacity shrink. In practice this becomes a computational heuristic: simulate flows under many scenarios, adjust edge capacities as if they were living tubes, and seek designs that remain functional across disturbances.

The most radical implication is psychological. Engineers are trained to distrust anything that looks like muddling through. A Steiner tree has the clean satisfaction of a theorem: it is provably minimal under its assumptions. A Physarum network looks like compromise, and compromise looks like failure if you are grading against perfection. But perfection is brittle when the world is noisy.

Biological algorithms do not outperform graph theory because they are more exact. They outperform it in the places where exactness is the wrong ambition. A living cell cannot afford to be exquisitely optimal for a single, frozen environment, because the environment will not stay frozen. Instead it uses rules that are locally sensible, globally stabilizing, and continuously revisable. That is distributed intelligence in an engineering register: no central planner, no exhaustive search of all possibilities, just a feedback loop that turns use into structure.

There is a quiet ethical dimension here as well. Minimal designs often hide their costs by pushing risk onto those who cannot absorb it. A neighborhood served by a single transit line is efficient on a map, until that line fails and the neighborhood is stranded. A hospital supplied by a single distributor is efficient until a disruption makes the savings look trivial beside the harm. Redundancy is not only a technical feature; it is a social promise that a system will keep serving people under strain. The slime mold, indifferent to politics but sensitive to survival, behaves as if that promise were non-negotiable.

A planner staring at a budget spreadsheet might still ask the hard question: how much redundancy is enough? Physarum cannot answer that in human terms. What it can offer is something subtler: a way to *generate* networks whose redundancies are not sprinkled uniformly but placed where flow and vulnerability make them valuable, and a way to adjust that placement as conditions change. The resulting maps do not look inevitable. They look alive, which may be closer to what infrastructure has to become in an era of shocks: not a static monument to an optimal past, but a system that can bend, reroute, and thicken its lifelines where the pressure truly falls.

8.2 Liquid Logic and Soft Machines

The appeal of a slime mold is not that it can be trained to perform party tricks, but that it forces a rethink of where "thinking" happens. In the familiar story, sensing is done by sensors, deciding is done by a processor, and acting is done by motors. The body is a vehicle for the brain. Physarum turns that story inside out. It senses with its surface, decides with its moving fluids, and acts by changing its own shape. There is no sharp boundary between computation and construction: the organism *computes* by becoming a different organism.

Engineers have a name for this style of problem-solving: *morphological computation*. The word "morphology" here simply means shape and structure. The claim is not mystical. It is practical: some of the work usually assigned to a control algorithm can be handed to the physics of a body that is built to respond in useful ways. A thermostat does not need to calculate the kinetic theory of gases; it only needs a bimetallic strip that bends predictably with temperature.

Morphological computation is the generalization of that insight from simple devices to complicated, adaptive machines.

The slime mold is an extreme case because it has almost nothing but morphology to work with. The control system is inseparable from the material. That makes it a surprisingly good teacher for robotics, where the limits of the conventional approach are easiest to see in the places that matter most: rubble, mud, crowds, oceans, human bodies. Environments like these are chaotic, and chaos punishes rigid plans.

A useful historical contrast comes from the early days of industrial automation. In mid-twentieth-century factories, robots were designed as precise machines in controlled spaces: hard floors, fixed parts, predictable timing. The classic industrial arm is a masterpiece of rigidity: stiff links, high-torque motors, careful calibration. It works because the world around it is made to behave. When engineers tried to carry that success into unstructured environments, the brittleness became obvious. A robot that can weld a car door all day can be defeated by a doorknob in a hallway if the knob is not exactly where it expects it to be.

Soft robotics emerged partly as a rebellion against that mismatch. Instead of treating compliance—the ability to deform—as an error to suppress, soft robotics treats compliance as a resource. A soft gripper can pick up objects it has never seen because it yields to their shapes. A soft crawler can traverse uneven ground because it conforms rather than collides. The body absorbs the complexity.

Physarum offers a deeper version of the same principle. It does not merely deform passively; it actively generates waves of contraction that push fluid through its own veins. The flow is not a by-product of movement. It is the movement.

Inside the plasmodium, cytoplasm shuttles back and forth through branching tubes in rhythmic pulses. These pulses are driven by biochemical oscillations that coordinate contraction and relaxation. The key engineering idea is that oscillators coupled through a shared body can produce coherent behavior without a central clock. When one region contracts, it forces fluid elsewhere. Because mass is conserved, local decisions are instantly entangled with global constraints. A change in one vein changes pressures throughout the network. In an organism, this coupling helps coordinate growth, transport, and reconfiguration. In a robot, the same coupling can coordinate locomotion, gripping, and adaptation.

This is the heart of "liquid logic": information carried not as bits in wires, but as pressures, flows, and rhythms in a deformable material.

To see why this counts as computation, it helps to loosen the usual definition. Computation, in the broadest sense, is a mapping from inputs to outputs according to rules. In a silicon computer the inputs are voltages, the rules are implemented by logic gates, and the outputs are voltages again. In Physarum the inputs might be chemical attractants, light, temperature, or mechanical touch. The rules are implemented by how flow reinforces tubes and how oscillations change under stimulation. The outputs are morphological: where the organism thickens, where it retreats, which tubes become highways and which shrink to threads.

Andrew Adamatzky popularized this perspective with the phrase *Physarum machines*. The point is not that you should replace your laptop with a petri dish. The point is that the organism can be *programmed in space*. Place attractants at certain locations and repellents at others, and the slime mold will form a pattern that reflects those constraints. The "program" is not a sequence of instructions. It is a landscape that shapes the organism's behavior.

That spatial programming idea has led to demonstrations of slime-mold logic, sometimes literal logic gates. If the term is unfamiliar, a logic gate is a tiny device that takes one or more binary inputs (0 or 1) and produces a binary output. The most basic gates are AND, OR, and NOT. Combine enough of them, and you can build any digital circuit, from an adding machine to a microprocessor.

Physarum logic gates exploit the fact that flows in a branching network can be steered. One path might be made attractive, another made hostile (for Physarum, light can serve as a repellent; certain chemicals can too). Mechanical stimulation can also modulate behavior: if a tube is gently pressed or disturbed, the streaming can change, sometimes temporarily blocking flow or diverting it. Adamatzky and Theresa Schubert (2014) demonstrated microfluidic-style gate designs in which living tubes functioned as reconfigurable channels, with the organism's own dynamics providing the switching.

There is a lovely inversion here. In electronics, the wires are passive and the switches do the work. In Physarum, the "wires" are alive and self-modifying, while the "switch" can be as simple as a patch of light or a physical obstruction. The circuit is not etched; it is grown. It can heal if damaged, reroute if blocked, and gradually rewrite itself as conditions change.

Calling this "logic" is accurate in a narrow sense, but the deeper contribution is more like a lesson in what kind of computation is natural to materials. Silicon is superb at fast, exact, repeatable operations. Living networks are superb at slow, approximate, context-sensitive operations that are robust to damage and noise. The mismatch is not a flaw. It is a clue about what each substrate is for.

Robotics sits right on that boundary. Many robots are asked to do tasks that resemble what animals do effortlessly: squeeze through a narrow opening, maintain grip on an object that shifts, crawl over loose debris, swim through turbulent water. In these settings, exactness is less valuable than grace under surprise. A controller that tries to calculate every contact force and every possible disturbance will drown in complexity. A body that naturally settles into useful behaviors can do much of the work without calculation.

Consider peristalsis, the wave-like squeezing motion used by your digestive tract to move food, or by earthworms to crawl. Peristalsis is not elegant in the way a legged gait can be elegant. It is relentless, adaptable, and hard to stop. A peristaltic robot does not need ankles, knees, or a precise foothold. It needs a body that can generate contraction waves and something like frictional asymmetry so that the wave produces net motion.

Physarum provides both a metaphor and a mechanism. The organism already produces contraction waves. It already uses friction with the substrate. It already routes fluid in a way that couples distant regions. A robot inspired by this does not copy the slime mold literally; it steals the principle: distributed oscillators linked by shared hydraulics can create coordinated motion.

A simple way to implement this is to build a soft tube with multiple chambers along its length, each chamber capable of contracting (perhaps pneumatically, hydraulically, or via shape-memory materials). If the chambers are connected by a common fluid line, then pressure changes in one chamber affect the others. Add local feedback—a sensor that responds to stretch or pressure—and the system can begin to act like a network of coupled oscillators. Instead of commanding each chamber with a centralized timing signal, you can let oscillations emerge. If one segment meets resistance (say, it presses against a rock), its deformation changes pressure, which shifts the phase of neighboring segments. The gait adapts without anyone "deciding" in a symbolic sense.

This is morphological computation in action: the control is performed by the interaction of material properties, geometry, and conservation laws. The robot "thinks through squishing."

A concrete anecdote makes the contrast with conventional control vivid. After the 2011 earthquake and tsunami in Japan, the Fukushima Daiichi nuclear disaster highlighted an uncomfortable truth: in hazardous environments, sending humans is dangerous, but sending robots is not straightforward either. Many robots that look impressive in a lab struggle with clutter, tight spaces, and unpredictable contact. Soft robots have been proposed for such tasks because a compliant body can squeeze through gaps, tolerate collisions, and avoid jamming where a rigid mechanism might get stuck or break. The dream is not a single heroic machine, but a class of devices that can enter dangerous spaces, adapt their shape to the situation, and continue functioning when parts are damaged.

Physarum-like control principles are attractive here because they do not require every contingency to be foreseen. A rubble field is an adversarial test for precomputed plans. It rewards systems that can generate behavior from local interactions. If a limb or segment fails, the remaining segments should not need a new software update to keep moving; they should naturally reorganize.

This is also why slime-mold-inspired thinking has begun to influence how engineers talk about *behavioral diversity*. In a traditional control mindset, switching among different patterns can look like instability. In a biological mindset, switching can be a feature: a repertoire of modes that the system can slip between as circumstances change. Physarum does not march steadily toward a goal with a single strategy; it explores, consolidates, and sometimes re-expands. That variability is not indecision. It is a way to keep options open in an uncertain world.

A soft robot designed in that spirit might not have one perfect gait.

It might have several mediocre gaits and a tendency to drift among them, with the environment nudging it toward whichever gait works best at the moment. The "controller" is less like a conductor and more like a set of local rules that keep the whole organism within a region of viable behavior.

None of this should be romanticized. The same features that make Physarum computation robust also make it frustrating from an engineering standpoint. It is slow. It is noisy. It is sensitive to uncontrolled details: humidity, substrate, temperature gradients, subtle chemical contamination. A living logic gate does not deliver the crisp, clocked certainty of a CMOS gate. Even in carefully designed experiments, accuracy can degrade when gates are cascaded into larger circuits. Errors accumulate. Signals drift. The organism's own tendency to remodel can undermine a carefully laid plan.

But those limitations also sharpen the lesson. Physarum is not competing with silicon at silicon's game. It suggests a different game: compute with matter when matter is what you must ultimately control. If the output you want is not a number but a shape, a path, a pattern of stiffness, then it can be more efficient to let physical processes do the heavy lifting.

There is an everyday analogy in how people learn to move. When you first learn to ride a bicycle, you might try to consciously compute balance: turn slightly left if you are falling right, and so on. It is hopeless. What actually happens is that your body learns a set of coupled reflexes through the physics of the bicycle itself. Balance is not calculated; it is entrained. The bicycle is part of the control loop. Morphological computation is a way of building machines that can be learned by their own physics, rather than commanded against it.

Physarum makes the same point in a more alien form. Its "intelligence" is distributed across a changing network. Its memory is partly in chemical states and partly in the thickness of tubes. Its logic is in the way flows reorganize when constraints change. If a robot is built from soft materials that already have rich dynamics, then forcing that robot to behave like a rigid machine with a central processor can be like trying to play a violin by hitting it with a hammer: you can get sound, but you are ignoring what the instrument is.

The promise of liquid logic is that it treats the body not as an inconvenient plant to be controlled, but as an active computational partner. A soft gripper that passively conforms is already doing morphology as computation. A soft robot whose internal hydraulics generate coordinated oscillations is doing even more. A system that can grow, heal, reroute flows, and redistribute effort when damaged edges closer to the slime mold's way of being: a machine whose control policy is written into its substance.

The future of this approach will not look like a petri dish bolted onto a chassis. It will look like new materials, new architectures, and new design habits: engineers thinking in terms of fields and flows, of local rules that yield global function, of machines that are not merely programmed but *cultivated.* When a robot can squeeze through a crack, maintain its grip without crushing, and keep moving after a puncture because its body automatically shifts the rhythm of its contractions, the computation is not hidden in a chip. It is in the way the machine changes shape, and in the way that change carries information forward.

8.3 The Limits of Intuition

A slime mold makes a persuasive salesperson. Put food at two points in a maze, and after a while a single thick tube remains between them. The image flatters a certain human hope: that intelligence is mostly a matter of *finding the right path*, and that nature, honed by evolution, will supply a cheap shortcut around difficult mathematics. It is hard to watch a living network prune itself into efficiency without feeling that something like wisdom is at work.

That feeling is not wrong, but it is incomplete. The same intuition that makes slime-mold experiments memorable can also make them misleading. Physarum is a powerful teacher when it comes to distributed problem-solving, yet it is also a reminder that a clever heuristic—a rule of thumb that works well in many cases—is not a guarantee. The organism does not carry a certificate of correctness. It carries a body, a set of feedback loops, and a history of what evolution asked it to survive. Engineers and scientists borrow from it at their own risk.

A useful way to think about these risks is to separate three kinds of limits. There are limits imposed by *geometry* (the shape of the environment can trap local behavior). There are limits imposed by *signals* (noise, conflicting cues, and missing information). And there are limits imposed by *logic* (the seductive step from "it often works" to "it will always work," especially when we try to chain small computations into larger ones).

The most visually obvious trap is a simple piece of geometry: the U-shape.

Imagine a U-shaped wall, like a horseshoe, open on one side. Place a food source just outside the open end, and put the slime mold on the far side of the wall. The organism expands, as it does, searching for chemical cues. If it touches the nearest part of the wall, it will tend to grow along it, because local gradients point that way and the physical surface provides an easy route. But the shortest route to the food is not the nearest wall. The shortest route requires a retreat: moving away from the food at first in order to find the opening of the U.

Humans recognize this kind of problem because it appears in everyday navigation. If you have ever followed your phone's map only to find a river between you and the destination, you have met a version of the same trap. The straight-line direction is not the direction that gets you there.

Physarum can be lured into the U. It is not "stupid"; it is following sensible local rules. The trouble is that a sensible local rule can be globally wrong. Many optimization problems, from route planning to machine learning, have an analogous pitfall called a *local minimum*. You can picture a landscape of hills and valleys where each position corresponds to a possible solution and the height corresponds to cost. If you always move downhill, you will end up in a valley, but not necessarily the deepest valley. The U-shaped trap is a physical metaphor for that abstract landscape. The slime mold is moving "downhill" in a chemical sense, climbing the wall that seems to lead to the reward, and it can stall.

Classical pathfinding algorithms have explicit strategies for this, like backtracking and systematic exploration. A program can be told: "If you reach a dead end, retreat until you find a junction and try another branch." Physarum does explore broadly at first, and sometimes that exploration saves it. Yet the very pruning that makes it impressive can also make it vulnerable. If the organism commits too early, reinforcing the wrong branch and starving alternatives, it can lock itself into a suboptimal configuration.

The engineer's lesson is not to mock the organism, but to name the condition under which it fails: when the environment rewards temporary retreat, local attraction can become a trap.

A second class of limits comes from the signals that guide growth. Physarum is often portrayed as following a single beacon: food. In a laboratory, that simplification is deliberate. The substrate is chosen to be uniform, lighting is controlled, temperature is stable, repellents are absent unless the experiment is explicitly about repellents. This is what makes the behavior legible. But the real world is a quarrel among stimuli.

Even in a petri dish, cues can conflict. A region might contain food but also light (often unpleasant for Physarum). Another region might be dark and humid but farther from nutrients. Some chemicals diffuse faster than others, creating gradients that are strong but misleading. If two food sources are present, their gradients interact. If the substrate has micro-texture, the organism may prefer certain paths for mechanical reasons that have nothing to do with transport efficiency.

When engineers borrow the slime mold as an algorithm, they often borrow the *cleaned* version: a neat world with a small set of nodes and a single objective. That makes sense as a first step, but it tempts a leap from demonstration to deployment. A planning problem in a city is not simply "connect A to B cheaply." It is "connect A to B while respecting land ownership, geology, politics, safety margins, and fairness." In supply chains it is "deliver reliably while facing changing costs, strikes, weather, and regulations." In power grids it is "maintain stability under uncertain demand and component failures."

Physarum's biological rules do not automatically incorporate those human constraints. They can be adapted into models that represent them, but the adaptation is the hard part. When a headline declares that slime molds design better networks than engineers, it is not merely exaggerating. It is erasing the place where engineering actually happens: the encoding of constraints, the choice of objectives, the negotiation of trade-offs.

Recent modeling work inspired by urban design has been unusually candid about this. When researchers try to turn slime-mold growth into a controllable design tool rather than a one-off demonstration, they run into missing mechanisms. Repellents and "anti-attractors" are harder to model robustly than attractants. Mixed stimuli can yield inconsistent outcomes. Some biological feedback loops are simplified or omitted, which may be acceptable for generating candidate networks but not for claiming faithful prediction of living behavior. The closer one tries to get to real design decisions, the more the details matter.

This is a familiar story in the history of bio-inspired engineering. Early aviation pioneers watched birds and learned essential principles about lift and stability, but airplanes do not flap like birds. They borrow from biology at the level of constraints and trade-offs, not at the level of copying. In the same way, slime-mold-inspired network design is most valuable as a way to explore a space of plausible solutions, not as a substitute for the messy reality of civil planning.

A third limit appears when intuition slides from biology into mathematics. The phrase "Physarum solves shortest paths" is both true and dangerously incomplete.

It is true in the sense that the organism, under certain controlled conditions, often leaves behind a thickened connection between food sources that is close to the shortest route. It is also true in a formal sense for specific mathematical models of Physarum-like adaptation. Vincenzo Bonifaci, Kurt Mehlhorn, and collaborators (2012) analyzed a model in which edge "conductance" increases with flow and decreases otherwise, and they proved convergence properties under the model's assumptions. That kind of result matters because it separates a charming anecdote from an algorithm with understood behavior.

But it is incomplete because the proof applies to the model, not to every petri dish and not to every environment. The model assumes a certain kind of continuous adaptation, a certain relationship between pressures and flows, and a certain stability of the system. Living

Physarum is not obliged to obey those assumptions perfectly. It is a biological entity with internal noise, biochemical thresholds, and competing drives. The proof draws a boundary around what is being claimed. Inside that boundary: convergence. Outside it: unknown.

This is not pedantry. It is exactly where intuitive stories become overconfident. Many scientific myths begin as a true observation that gets generalized beyond its domain. "It solved a maze once" becomes "it can solve mazes." "It made a network resembling Tokyo" becomes "it can plan cities." The risky step is not in the first statement; it is in the second.

For engineers, the phrase *under the model assumptions* is not a disclaimer to be mumbled and ignored. It is the instruction manual. If a Physarum-inspired algorithm is implemented in software for a logistics problem, the software must define what corresponds to conductance, what corresponds to flow, and what corresponds to reinforcement. If the system's real constraints break the analogy, the algorithm may behave in surprising ways. Sometimes the surprise is good; sometimes it is catastrophic.

The composability problem—what happens when you chain many small pieces into a larger system—is where these limits become brutally quantifiable. A single logic gate made from Physarum is a charming demonstration. But digital computation depends on building large circuits by wiring gates together. Silicon works because the gates are reliable, their outputs are standardized, and they can be cascaded indefinitely. Living substrates do not offer that guarantee.

In Physarum-based logic experiments, gate behavior can be probabilistic, sensitive to timing, and vulnerable to drift. A gate that works 95% of the time is already problematic; when you combine many such gates, the probability that the entire circuit works can collapse. Even if individual accuracies seem respectable, errors multiply. The organism's own tendency to remodel its tubes, which is a virtue for robustness in a physical environment, becomes a vice for stable computation. A tube that strengthens because it carries flow might unintentionally create a shortcut that changes the intended logic. A temporary perturbation can have long-lasting structural consequences.

This is a different kind of limitation from the U-shaped trap. It is not a failure of local guidance; it is a failure of *standardization*. Digital engineering thrives on components that behave the same way regardless of context. Biology thrives on components that change behavior with context. Physarum is excellent at adapting; adaptation is what makes it hard to treat as a fixed gate.

The lesson is not that living computation is futile. It is that living computation excels when the output is itself adaptive: when you want a system that can reconfigure, heal, and respond in parallel, rather than a system that must execute a long sequence of exact, brittle steps. A slime mold is a bad calculator and a good planner, if "planner" means an entity that can negotiate trade-offs on the fly.

Even in the network-design setting, where Physarum has its most famous successes, intuition can be trapped by scale. A petri dish network might involve a dozen nodes. A real power grid has thousands of buses and lines, and it must meet stability constraints that are not captured by simple distance metrics. A supply chain spans jurisdictions, tariffs, contracts, and shifting consumer behavior. A city transport system is embedded in land use and human habits. The more complex the system, the more a naive mapping between "food" and "demand" will break.

Bio-inspired models respond by abstracting. They take the core transferable rule—reinforce edges that carry flow, prune those that do not—and they apply it in a simulated domain where additional constraints can be encoded. That is a perfectly respectable engineering strategy, and it is often where the true value lies. The organism provides not an answer but a *design principle*. The danger is mistaking the source of the value. The value is not that a slime mold is secretly a civil engineer; it is that a simple feedback rule can generate networks that occupy a useful compromise between cost and resilience.

In that sense, the slime mold is like a well-chosen metaphor that happens to be alive. It embodies an algorithm in flesh, which makes the algorithm harder to ignore and easier to remember. But metaphors have failure modes. They hide assumptions, and they can seduce the imagination into thinking that the world is simpler than it is.

One of the most instructive moments in any slime-mold experiment is the time before the final answer appears. The organism spreads out in a messy, almost indecent abundance. It does not behave like a tidy optimizer marching toward a solution. It behaves like a system hedging its bets, sampling possibilities, laying down provisional links. Only later does it prune. That two-stage behavior is often the secret behind its apparent cleverness. The early sprawl protects it from traps; the later pruning protects it from waste.

But even that strategy has limits. If the sprawl is too timid, the organism misses the opening of the U. If the pruning is too aggressive, it commits too early and cannot recover. If the environment is noisy, the sprawl may be pulled in too many directions and never consolidate. If the environment changes quickly, the organism may be perpetually remodeling, never settling into a stable network.

These are not failures to be explained away. They are the boundaries of a particular form of intelligence: distributed, material, local, and adaptive. Within those boundaries, Physarum's heuristics can outperform our pristine mathematical ideals precisely because the ideals ignore the kinds of stress real systems face. Outside those boundaries, the heuristics can fail in ways that are not mysterious at all once the underlying constraints are named.

A reliable engineering relationship with biology begins at that point: not when a living system dazzles us, but when it disappoints us in a way we can understand. A U-shaped trap is a gift because it reveals what the organism cannot do without help. A drifting logic gate is a gift because it reveals what “computation” means when the substrate is alive. A model proof bounded by assumptions is a gift because it shows exactly where formal guarantees end and empirical exploration begins.

The slime mold remains a valuable collaborator, not because it is always right, but because it makes it hard to forget that the world is full of trade-offs. Efficiency without slack is fragility. Redundancy without guidance is waste. Local rules can yield global order, and they can also yield global mistakes. The real sophistication lies in learning when to trust the organism’s instincts, when to temper them with classical control, and when to change the environment so that a simple heuristic becomes the right tool for the job.

Chapter 9

Redefining Mind

If a slime mold can navigate a maze without a brain, does it possess a mind? This chapter challenges the neurocentric bias of biology, arguing that intelligence is not a privilege of the cortex but a fundamental property of organized matter. By re-examining how we measure and define thought, we discover a spectrum of sentience that extends far beyond the animal kingdom.

9.1 The Lens of Time

A slime mold offers a particular kind of insult to human common sense. You place a smear of yellow living gel on a damp surface, give it a few oat flakes, and come back expecting movement—some sign of purpose, a twitch, an advance, anything that fits the animal template. Instead you see what looks like yesterday's spill: a quiet, irregular patch, perhaps a little larger, perhaps not. If you did not know it was alive, you might file it mentally with lichen and rust. Even knowing better, the temptation is to treat it as a passive substance reacting to chemistry the way litmus reacts to acid. The puzzle is not that slime molds behave strangely. The puzzle is that they behave at all, and yet do so at a pace and scale that slip past our ordinary sensors.

Human beings are fast-minded mammals. We live in a world where the difference between a mistake and a meal can be a fraction of a second. Our nervous system is tuned to changes that happen in the tempo of footsteps, thrown objects, facial expressions, and speech. When something changes more slowly than a houseplant, we start to treat the change as background, like the weathering of stone. The slime mold *Physarum polycephalum* is not immobile; it is simply operating on a timescale that makes it *look* immobile. Intelligence, in practice, is often something we decide we have witnessed. Before the slime mold can be accused of lacking a mind, it has to be granted the courtesy of being seen.

The key trick is not a new theory of cognition but a new relationship to time. Time-lapse macro photography does something almost philosophical: it translates hours into seconds and turns a barely perceptible creep into an unmistakable act. Under that compression, what appeared as a smear becomes a restless network, probing, pausing, retreating, and committing. Tubes thicken in some directions and wither in others. Branches are sent out like exploratory fingers and then retracted, as if the organism had changed its mind. If you have ever watched a time-lapse of clouds boiling over mountains, you know the uncanny effect: you are not seeing a different world, only the same world played back at a tempo that reveals its hidden dynamics. With slime molds the effect is sharper, because what is revealed looks like decision.

The reliance on imaging is not merely a matter of convenience. It is closer to a lesson about the observer. Many of the behaviors we call cognitive are recognized through patterns of change: approach and avoidance, trial and error, persistence, learning. If the changes unfold too slowly for our unaided senses, we will systematically underestimate the agent. A slime mold is a rebuke to the idea that intelligence is a bright inner light that shines regardless of who is looking. Intelligence is often inferred from *trajectories*—from the way an organism moves through possibilities over time. Without the right lens, those trajectories vanish.

A historical analogy helps. For centuries, the night sky was a catalogue of fixed points and a handful of wandering lights. The planets were called wanderers because they refused to stay put, but even their wandering was stately enough to track with the naked eye. It took the telescope to reveal moons circling Jupiter, and it took repeated observation to make those moons persuasive evidence that not everything orbited Earth. The telescope did not create the moons; it created the *visibility* of a pattern that forced a revision of assumptions. Time-lapse is a telescope for tempo. It makes a case, frame by frame, that slow is not the same as simple.

One concrete demonstration of this shift occurred when researchers began placing *Physarum* in experimental mazes with food at two points. The setup is disarmingly straightforward: a labyrinth cut into agar, the slime mold introduced at one entrance with a food source, another food source placed at the exit. The organism spreads through the channels, then, over time, abandons redundant routes and consolidates a single thickened tube connecting the food sources. Watched in real time, the experiment is mostly waiting punctuated by mild disappointment: the maze seems unchanged for long stretches. Watched in time-lapse, it becomes an argument about computation embodied in flesh. The organism explores, compares, and then commits to an efficient path. Even if one resists the word *solve*, it is hard to avoid the sense that the organism is doing something more than drifting.

The fascination here is not simply that the slime mold finds a short path. It is that the process resembles what many people imagine a mind does: consider options, test them, prune them, and keep what works. A human solves a maze by running a model in the head, perhaps imagining turns and dead ends. The slime mold solves a maze by remodeling itself. The algorithm is written in growth and retraction, in thickening and thinning, in the strengthening of one tube at the expense of another. But the recognition of an algorithm depends on seeing the *sequence*. If you only see the beginning and the end, you can always tell yourself a story in which the outcome is accidental. Time-lapse supplies the missing middle—the evidence of search.

This need for temporal translation is not unique to slime molds. Microbiologists have long relied on time-lapse microscopy to make sense of bacteria that divide, swarm, and coordinate. Developmental biologists watch embryos by compressing hours of cell division into minutes. In each case, the imaging is not an ornament; it is a way of matching our perceptual bandwidth to the organism's. A useful phrase in this context is *behavioral resolution*. Just as a camera needs enough spatial resolution to distinguish a face from a blur, an observer needs enough temporal resolution to distinguish a decision from noise. Too slow a sampling and you miss the turns; too fast and you drown in detail.

Macro time-lapse of *Physarum* is often filmed at intervals of tens of seconds to minutes. That choice is not arbitrary. The organism's most dramatic visible changes—the advance of a fan-shaped front, the retreat from an unfavorable region, the reinforcement of a particular tube—unfold over minutes and hours. Compressing time makes those changes legible. But it also creates an illusion that can mislead: the slime mold seems to move continuously, like an animal, while in reality its motion is produced by rhythmic internal flows. The organism's body is a living network of tubes filled with flowing cytoplasm, and the flow oscillates back and forth. At one moment fluid streams in one direction; a minute later it reverses. Over many cycles the imbalance in those oscillations, combined with growth at the leading edge and shrinkage elsewhere, yields net movement. Time-lapse hides the oscillation and reveals the outcome. For understanding, both views matter: the slowed-down motion is how we recognize purpose, while the close-up oscillation is how we learn the mechanism by which purpose is carried out.

Here the lens of time does something more subtle than speed things up. It encourages a change in what counts as a meaningful act. When a dog decides to turn left, the decision looks like a discrete event: the body pivots, the paws change direction, the animal commits. With *Physarum*, the equivalent decision is distributed across a field of tiny changes: a branch grows a little faster in one corridor, a tube elsewhere becomes slightly less used, a nutrient gradient nudges the balance of flow. There is no single pivot. The decision is spread out, and it is made by the collective dynamics of the network. A human observer without time compression sees only a slowly shifting puddle. A time-lapse observer sees a pattern of distributed commitments.

One of the most striking stories in this area comes from experiments on temporal expectation. The general idea is simple enough to tell at a dinner party: can a slime mold, with no neurons, learn the timing of a recurring event? Researchers exposed *Physarum* to periodic unfavorable conditions—for example, a brief spell of dryness or a mild chemical stress—at regular intervals. After a few cycles the organism began to slow its movement *in advance* of the expected stress, even when the stress was omitted. The behavior looks uncannily like anticipation. But anticipation is not a single snapshot; it is a relationship between what happens now and what is expected later. To recognize it, one needs a record that spans hours and a way of aligning events across time. Time-lapse and careful tracking become not only observational tools but conceptual tools: they allow the question "Does it prepare?" to be asked in a form that can be answered.

This is where time manipulation shades into an argument about what a mind is allowed to be. A traditional picture of cognition assumes quick internal processing: sensory inputs arrive, the brain computes, the body responds. *Physarum* invites a different picture. Its processing is slow, but it is also continuous, like a river adjusting its course. Its *memory* is often not a stored symbol but a persistent change in its own structure or chemistry. Its *choices* are not the flip of a neural switch but the gradual reweighting of flows through a network. If you insist that cognition must be quick and punctuated, the slime mold will always look mindless. If you allow cognition to be slow and morphological—written into shape and dynamics—a different class of intelligence comes into view.

The emphasis on time has a practical consequence: it tells you which experiments are even possible. A creature that moves millimeters per hour cannot be studied by the same methods used for rodents. You cannot present a stimulus for half a second and expect a response. You cannot interpret a brief pause as indecision. Instead, you design questions that match the organism's cadence. You ask about gradual preference: does it grow more toward one chemical than another over an hour? You ask about history dependence: does yesterday's exposure change today's path? You ask about the shape of a network under steady conditions, and how that shape changes when conditions shift.

When scientists build these experiments, they often discover that the most informative *behavior* is not locomotion but *reconfiguration*. The slime mold's body is already a map of its past: thick tubes mark well-traveled routes, while thin or abandoned tubes mark paths that were tried and then rejected. Under time-lapse, the organism becomes a living record of its own exploration. That record is not hidden in a skull. It is out in the open, visible as geometry. There is a kind of honesty to it: the organism cannot conceal its deliberations because it must embody them.

A vivid example involves what has been called externalized memory. *Physarum* leaves behind a slime sheath as it moves, a thin trail that can influence later behavior. In certain experimental setups the organism avoids its own previous paths, effectively using the trail as a cue to prevent redundant exploration. One can tell a mundane story: perhaps the trail is simply repellent, and avoidance is reflex. But time-lapse reveals something more structured. The organism explores a space, lays down traces, and later navigates as if those traces are a spatial memory of where it has been. Whether one describes this as "memory" depends partly on philosophical taste, but the empirical point is clear: the organism's past is written into the environment in a way that shapes its future. Again, the phenomenon is fundamentally temporal. A still photograph shows only a smear. A time series shows a history.

If this sounds like we are smuggling human concepts into a blob of protoplasm, it helps to remember that humans have always relied on instruments to extend perception, and those instruments have repeatedly forced revisions of what counts as action. Consider the slow world of plants. Long before slime molds became a laboratory celebrity, Darwin and his son Francis studied plant movement, including how seedlings bend toward light and how climbing plants circle as they search for support. Many people assumed plants were inert, yet careful observation showed purposeful-seeming motion, just at a pace too slow for casual notice. Darwin treated these movements as behavior deserving of explanation. Modern time-lapse makes that behavior obvious in a way that nineteenth-century observers had to labor to demonstrate. Slime molds occupy a similar space in our imagination. They look inert until the right apparatus insists otherwise.

The apparatus matters aesthetically as well as scientifically. Watch enough time-lapse of *Physarum* and you start to feel that the organism has moods. It hesitates at boundaries. It surges toward food. It seems to "test" a route with a thin probe before sending more of itself. These are, strictly speaking, anthropomorphic impressions. But the impressions are not arbitrary; they are our brain's way of compressing a complex temporal pattern into a readable narrative. The danger is taking the narrative too literally. The value is that it encourages questions: what physical process corresponds to hesitation? what internal variable corresponds to preference? The lens of time does not merely reveal a hidden truth; it creates the conditions under which curiosity becomes precise.

There is another twist. Changing the playback speed changes not only what we see but what we think is causally important. At high speed, the slime mold's movement looks smooth and directed, and one might imagine a central controller. At more natural speed, one becomes aware of how often the organism changes direction, how much it explores, how much "waste" there is in the form of abandoned branches. That waste is not a failure; it is a method. Many intelligent systems explore broadly, then compress. A mind that never wastes effort is usually a mind that never learns. The slime mold's time-lapse behavior teaches a general principle: intelligence often looks inefficient in the moment and efficient in retrospect.

The lens of time also complicates the boundary between organism and environment. A fast-moving animal carries most of its structure with it. A slow-growing network, by contrast, reshapes the world it inhabits—drying a patch by absorbing moisture, consuming nutrients, leaving trails, changing microbial communities—and then responds to those changes. Over hours, the environment becomes part of the organism's effective body. When you watch a time-lapse, you are not just watching a creature move through a static maze; you are watching a creature and a landscape co-evolve. This can make the organism appear more intelligent than it would in a perfectly passive environment, but that is precisely the point. Intelligence is often a property of an organism embedded in a world, not an isolated brain in a jar.

A final reason time-lapse matters is methodological humility. In the early days of slime mold experiments, it was easy to overinterpret results by focusing on dramatic endpoints: a neat tube connecting food sources, a network resembling a human-designed system. With better imaging and longer recordings, researchers could see how variable the process is. The organism may try many routes before settling. It may settle differently depending on humidity, temperature, and minute differences in the substrate. It may "change its mind" if the environment changes. Time-lapse gives access to variability, and variability is where mechanism hides. A single perfect run looks like magic; many runs with their hesitations and detours look like biology.

There is a temptation, when confronted with a slow intelligence, to treat it as a novelty: a party trick of evolution. Yet the deeper implication is that our concept of mind has been trained on the creatures most like us. We built our theories around fast animals with eyes and limbs, then treated those theories as definitions. Slime molds expose the provincialism of that approach. If intelligence is recognized through behavior, and behavior is recognized through change, then the timescale of change becomes a gatekeeper. The gatekeeper can be moved.

One could say that time-lapse makes the slime mold intelligent by making it look intentional. But that gives the camera too much credit and the organism too little. The camera does not invent the pattern; it reveals it. What it does invent is our *access* to the pattern. That access has consequences for how questions are framed, how experiments are designed, and how boldly we are willing to use words like learning, memory, and decision outside the familiar territory of brains.

The slime mold, under the lens of time, becomes a reminder that cognition is not only an inner drama. It can be a visible rearrangement, a slow inscription of preference into structure, a living record of exploration written in tubes and trails. When hours are folded into seconds, the organism stops looking like a stain and starts looking like an agent—not because it has become something else, but because, at last, it is moving at a speed we can understand.

9.2 Cognition in the Concrete

Once the eye has been trained to see *Physarum polycephalum* as an active thing rather than a stain, a more uncomfortable question follows. If it can make what look like sensible choices—finding food, avoiding hazards, pruning wasteful routes—what is doing the choosing? The reflex answer, drilled into us by a century of neuroscience, is that choice requires neurons: electrical spikes, synapses that strengthen and weaken, networks that represent the world. A slime mold has none of that. It is a single giant cell, spread out into a lacework of tubes, with many nuclei sharing the same cytoplasm. Its "brain" is everywhere and nowhere, because its information processing is not carried by discrete messages passed between specialized cells. It is carried by matter moving through matter.

The temptation is to say: then it is not really processing information; it is merely obeying physics. But neurons obey physics too. A nerve impulse is an electrochemical wave; a synapse is a molecular machine. The real difference is not physics versus something magical. The difference is *which* physical substrate does the work. In *Physarum*, the substrate is hydrodynamic: flows of cytoplasm, oscillating pressures, and a living plumbing system that remodels itself in response to use. If neural cognition is often told as a story of spikes and circuits, slime-mold cognition is a story of pumping and pipes.

A few terms help, and they are wonderfully literal. The slime mold's body is threaded with tubes that carry cytoplasm. Inside these tubes the cytoplasm performs what biologists call *shuttle streaming*: a back-and-forth flow driven by rhythmic contractions of the cell's actin-myosin machinery, rather like the molecular cousin of muscle. The flow reverses direction on a cycle of roughly a minute or two. Nothing about that sounds like thinking, until one notices what the flow accomplishes. It distributes nutrients and chemical signals. It couples distant parts of the body so that a change at one edge can influence behavior elsewhere. Most importantly, it creates a feedback loop between *where the fluid goes* and *what the tubes become*.

A tube that carries more flow tends to thicken; a tube that carries less flow tends to thin and sometimes vanish. That single rule, repeated across a branching network, is already a kind of memory. Not memory in the sense of an image stored in a mental drawer, but memory as a durable change in structure that reflects past events. A thick tube is a fossil of yesterday's traffic. A thin, fading tube is a record of a route that stopped paying off.

This is why the slime mold can look uncannily like it is running an algorithm. Toshiyuki Nakagaki and colleagues (2000) made the point famous with their maze experiments: the organism initially spreads into many corridors, then gradually retracts from dead ends and redundant paths, leaving a stronger, shorter connection between food sources. The drama is not a sudden insight. The drama is that, at each moment, local dynamics accumulate into a global solution. The organism does not need to "know" the maze in advance. It only needs to keep reshaping itself in a way that rewards successful transport.

To make this less abstract, imagine a city that has no traffic lights, no central planning office, and no maps, but where every road physically changes shape depending on how many cars use it. Busy roads widen; empty roads narrow; a rarely used bridge might quietly disappear. After enough time, you would end up with a network that reflects the city's habits: efficient routes between important places, fewer redundant side streets, and multiple paths where reliability matters. That is close to what *Physarum* is doing, except that its "cars" are cytoplasm and its "roads" are living tubes.

Atsushi Tero and colleagues (2010) took this idea and turned it into a mathematical model that became widely cited well beyond slime-mold biology. They described adaptive transport networks whose links thicken with flow and decay without it, capturing how *Physarum* balances competing demands: short paths that save material, redundant paths that preserve resilience when something breaks, and enough connectivity to keep supplies moving. The reason engineers paid attention is simple: these are the same trade-offs faced by rail systems, power grids, and supply chains. When a slime mold spreads across a map-like arrangement of food sources, the resulting network can resemble human infrastructure not because it is copying us, but because both are constrained by geometry and cost.

The interesting step, for a theory of mind, is to notice that the physical rule "flow reinforces tube" does more than maintain transport. It *computes* in the sense that it transforms information about the environment into a structured response. Nutrients and hazards alter local physiology; local physiology alters flow; flow alters tube thickness; tube thickness alters future flow. The organism becomes a material feedback system that turns experience into architecture. A neural network changes synaptic weights; a slime mold changes diameters.

The phrase "hydrodynamic logic" can sound like overreach, as if one were claiming that a tube is literally a thought. The safer claim is that certain patterns of flow and remodeling function like logic gates in a circuit: they select, amplify, and suppress pathways in a way that implements decisions. Consider a branching point where the slime mold can send resources left or right. If the left branch leads to food, chemical cues and improved energy supply tend to increase oscillatory activity and growth on that side, drawing more cytoplasm into the left branch during streaming. Increased flow then thickens the left tube, which lowers resistance and makes it even easier for flow to pass that way. The system behaves like a positive feedback loop: success breeds reinforcement.

Now consider the opposite case: an aversive chemical or a dry region on the left. The slime mold may reduce growth there, contract more strongly, or change the rhythmic pattern of streaming so that less cytoplasm enters that branch. Reduced flow leads to thinning, which increases resistance, which further reduces flow. That is a negative selection: failure breeds withdrawal. No neuron has fired, but a choice has been enacted in the only language the organism speaks fluently: pressure, resistance, and shape.

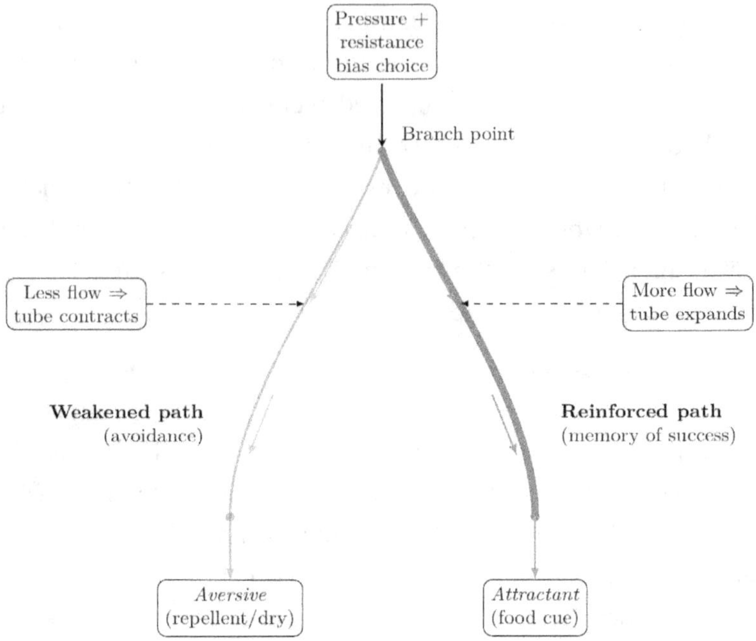

The diagram is schematic, but it captures a crucial shift in intuition. With neurons, one often imagines information as something that travels while the wiring stays fixed. With *Physarum*, the wiring is the message. The organism computes by *becoming* a particular network.

The same physical loop helps make sense of learning phenomena that seem, at first glance, to belong exclusively to animals. Habituation is a classic example: an organism gradually reduces its response to a repeated, harmless stimulus, and then the response can recover after a rest. When Romain Boisseau, David Vogel, and colleagues (2016) demonstrated habituation-like behavior in *Physarum*, they were careful to use mild chemicals that were unpleasant but not damaging. Repeated exposures led the slime mold to become less avoidant. That looks like the simplest form of learning: adjusting behavior based on experience.

A neurocentric story would search for a cellular analog of synaptic plasticity. The slime-mold story looks different. Repeated exposure can alter the organism's internal chemical state, its contractile rhythm, or the sensitivity of its signaling pathways. Those changes then shift streaming patterns and growth decisions. The memory is not stored in a dedicated module; it is distributed across the cell's physiology. In a system like this, learning does not have to be a special cognitive layer added on top of metabolism. It can be a change in the physical parameters that govern the flow-shape feedback loop.

This is also why *Physarum* can appear oddly "reasonable" in complex environments without requiring a hidden controller. Chris Reid and Tanya Latty (2012) showed that the slime mold can use its own slime trail as a cue to avoid previously explored areas. In maze-like traps, this reduces wasted exploration and improves navigation. The word *memory* is not a metaphor here; it is a functional description. The organism leaves a trace, and the trace alters future decisions. What is striking is how little machinery is required. The environment becomes part of the information-processing apparatus. The slime mold does not need to carry a detailed internal map if it can write notes onto the world and later read them.

A human analogy makes the point. When you park in a giant lot, you might "offload" memory by taking a photo of the nearest sign, or by dropping a pin on a map. That does not make you less intelligent; it makes you pragmatically intelligent. *Physarum*'s slime trail is not a smartphone, of course, but it plays a similar role: an external record that stabilizes behavior over time. The cognition is not confined within a boundary. It is smeared, literally, across the surface.

Once one begins thinking in these terms, the maze experiment looks less like a parlor trick and more like a consequence of a general design principle: when a system must transport resources through a changing environment, it can "think" by continually adapting its transport network. The model is not a picture in a head; it is a pattern in pipes. A short path emerges because short paths carry flow efficiently. Redundant links persist when they protect against disruption. Dead ends are pruned because they drain resources without payoff. In animals, similar pressures shape nervous systems: frequently used pathways strengthen; unused ones weaken. The difference is the currency. Neurons trade in spikes; slime molds trade in pressure and cytoplasm.

This concrete view of cognition also makes the slime mold's limitations more interesting rather than less. The organism is not omniscient. It can be tricked by conflicting cues. It can get stuck if the physical substrate prevents exploration. It sometimes commits too early or too late. But those are exactly the failure modes one expects from a system that computes by local feedback rather than centralized planning. The point is not to crown it as a miniature genius. The point is to see intelligence as something that can be built out of many kinds of parts, provided those parts support certain loops: sensing that changes internal state, internal state that shapes action, action that changes the conditions for future sensing.

That loop is not an abstract slogan. In *Physarum* it has a palpable feel. Put an oat flake at one edge of a Petri dish and you can watch the organism thicken the tubes that serve that region, like a city quietly upgrading a road that leads to a profitable port. Introduce a mild repellent and you can watch the network reroute, not by picking up and walking away but by slowly starving one set of tubes and feeding another. Over time, the organism turns the world into gradients it can exploit and architectures it can inhabit. It is hard to watch this without sensing a kind of stubborn purposiveness: a determination to maintain flow, to keep contact with food, to remain viable in a hostile geometry.

There is a small, almost comic, human moment embedded in many of these experiments: the need to wait. Researchers who work with *Physarum* develop patience not as a virtue but as a technique. You inoculate a dish, set up your gradients, and then you go home. The computation runs overnight. In the morning the organism has produced an answer in the only form it knows: a changed body. That schedule is a reminder that cognition is not always the flash of a solution. Sometimes it is the slow accumulation of physical commitments.

The lure of the slime mold is that it offers a stripped-down case. It removes the familiar scaffolding of brains, muscles, eyes, and limbs, leaving something that still manages to integrate information and act coherently. The explanation does not require a ghostly mind inside the cell. It requires attention to how living matter can store history in its own shape, how flows can coordinate distant regions without a central dispatcher, and how the world itself can serve as scratch paper for a creature that is willing to write with its body.

When the slime mold thickens one corridor and abandons another, it is not sending a message to itself saying "this way." It is *making* "this way" easier to be true the next time fluid surges through. In that sense, memory and decision are not separate stages. They are two views of the same physical act: reinforcing what worked, withdrawing from what did not, and letting the next moment inherit the biases of the last.

9.3 The Distributed Self

A familiar picture of intelligence has a familiar shape: a head. Even when the head is not literally a head, it is a central hub—a brain, a controller, a command center—that collects information and issues decisions. The rest of the body follows orders. This picture is so deeply built into everyday language that it can feel like common sense rather than a theory. "Use your head." "Make up your mind." "Think it through." The metaphors point upward, inward, toward a single place where the real work supposedly happens.

A slime mold makes that geography hard to sustain. Its activity is all frontage: sensing at the edges, resources moving through an expanding and contracting network, yesterday's successes preserved as thickened tubes. The organism behaves as a coherent agent, yet there is no privileged site where the coherence lives. It can be cut, and the pieces may continue to function and later fuse. It can be stretched thin, and still coordinate. It can reshape itself around obstacles without any part being in charge. If a self exists here, it is not an organ. It is an organization.

That shift—from self-as-place to self-as-pattern—is more than a clever philosophical maneuver. It changes what counts as an intelligent system in nature and in technology. It also changes how fragile we should think a self is. The standard worry about decentralized systems is that they will fall apart without leadership. The slime mold suggests an opposite worry: a system may remain stubbornly itself even when there is no leader to remove.

A useful starting point is to ask what is meant by *self* in the first place. In everyday life, a self is the thing that persists when conditions change: the one who keeps promises, remembers where the keys are, recoils from pain, recognizes a name. In biology, persistence has a blunt requirement: stay alive. Living systems must keep certain internal variables within viable ranges—water content, salt balance, energy supply—while the environment tries to push them out of range. That continual act of keeping-within-bounds is not glamorous, but it is a kind of goal-directedness, and it comes before brains in evolutionary history. A bacterium swimming up a sugar gradient is not reflecting on dessert; it is maintaining a metabolism. Yet the behavior has all the minimal ingredients that later become associated with mind: sensing, preference, action, and adjustment.

The slime mold sits in a sweet spot for thinking about this. It is large enough to show rich behavior and simple enough to make it difficult to hide behind the usual explanatory crutches. When it forms an efficient transport network between food sources, it is doing something that looks like planning. When it prunes redundant tubes, it is doing something that looks like commitment. When it avoids its own trails, it is doing something that looks like memory extended into the world. The challenge is to say what holds all this together as *one* agent.

One answer, common in engineering, is to treat the self as a *control system*. A thermostat is not conscious, but it does have a boundary and a goal: keep temperature near a set point. It senses, compares, and acts. The problem with that analogy is not that it is too simple; the problem is that it is too centralized. The thermostat has a sensor and a switch; the world outside is something it acts upon. *Physarum* is not like that. Its sensing is distributed. Its actuation is distributed. Even the "set points" are distributed: gradients of chemical cues and hydration become embodied preferences across the organism's surface.

Another answer comes from contemporary theoretical biology and cognitive science: the self can be defined by the way a system partitions the world into *inside* and *outside*. The technical language is often that of *Markov blankets*, a term popular in discussions of active inference and hierarchical self-organization, including work by Palacios, Razi and collaborators (2020). The basic intuition can be put plainly. To be a system at all, you must have some boundary that mediates your exchange with the world. You do not have to be sealed off; in fact, living systems cannot be sealed off. But you do need an interface: something that is acted on by the environment and that, in turn, acts on the environment. Skin, membranes, sensory surfaces, chemical receptors, even patterns of behavior can serve this role.

The appeal of this boundary-focused view is that it does not assume a central headquarters. It makes the self a matter of *coupling*: which changes in the world reliably influence which internal changes, and which internal changes reliably influence the world. The boundary is where information becomes action and action becomes information again. A slime mold has a boundary in the most literal sense—its outer membrane—but it also has functional boundaries that shift as it grows and retracts. In a branching network, "inside" is not a fixed volume. It is a moving territory, constantly re-drawn.

There is a further twist, and it matters for what kinds of selves are possible. Boundaries can be nested. A multicellular organism is a boundary made of smaller bounded things (cells), and those cells contain bounded components (organelles). In social animals, the effective boundary of problem-solving can extend beyond the skin: a group might remember collectively where resources are, coordinate defense, and make decisions no single member could make alone. That possibility makes the self less like a dot on a map and more like a set of overlapping circles that can expand and contract.

Michael Levin's work on bioelectricity and scale-free cognition (2019) has pushed this idea in an especially provocative direction. Bioelectric signals—slow voltage patterns across tissues, carried by ion channels and gap junctions—help coordinate development and regeneration in animals, guiding cells toward large-scale goals like "build an eye here" or "repair the limb." Those goals are not held in a single neuron. They are distributed across networks of cells. The implication is not that every tissue is a brain, but that goal-directed organization can appear at multiple scales, depending on what channels of communication and coordination are available. A self, on this view, is not a privilege granted by the presence of a cortex. It is a property that can be negotiated by biological networks.

The slime mold offers a biological parable of the same point. Its coordination channel is not bioelectric patterning across tissues (though electrical signals and ion dynamics matter in many cells), but rhythmic flows and mechanical tensions that bind distant regions into a single functional unit. The organism's "decisions" emerge from local updates, yet the result is coherent at the whole-network level. That coherence is not an illusion; it has consequences. Resources end up where they are needed. Risks are avoided. The body is maintained and reorganized in ways that keep the organism viable.

A concrete example keeps this from floating away into abstraction: the slime mold's transport networks that echo human infrastructure. When *Physarum* is presented with food sources arranged like cities, it can form a network that balances efficiency and robustness, sometimes resembling rail networks such as those around Tokyo, which helped make the phenomenon famous. What matters is not the publicity value of a biological "engineer." What matters is the kind of agency on display. A rail network is not the product of a single mind, even when it has planners. It is the product of many constraints, negotiations, historical accidents, and feedback from use. A slime mold network is similarly the product of constraints and feedback, but the feedback is embodied in the organism's own material. The slime mold is both the builder and the built.

That looping identity has a psychological echo that is easy to miss. Humans often treat the self as something that owns the body, like a driver in a car. Yet much of what gives us a stable identity is not a conscious driver but habits, hormonal cycles, immune responses, and learned patterns that run below awareness. A distributed self is not alien to us; it is simply less flattering. The slime mold forces the point by stripping away the flattering parts.

Philosophers of mind have argued that cognition can sometimes extend beyond the skull. Andy Clark and David Chalmers' "extended mind" proposal (1998) is often introduced with a simple story: if a notebook reliably functions as a person's memory store, used automatically and constantly, then the notebook is not merely a tool; it is part of the cognitive system. Whether one agrees with the conclusion, the example is hard to forget, because it highlights a practical truth. People think with calendars, calculators, maps, and other people. We distribute cognition across artifacts and social arrangements all the time.

Distributed cognition research, associated with thinkers like Edwin Hutchins, made the case empirically by studying real-world tasks such as ship navigation: the "mind" of the navigating system is not located in any one sailor's head but in the coordinated dance among people, instruments, checklists, and routines that unfold across time. The unit that solves the problem is a socio-material system. No single component contains the whole solution.

Slime molds provide a biological cousin of this phenomenon. When *Physarum* lays down a slime trail and later uses it to avoid revisiting areas, the environment is not a passive stage. It is part of the memory system. The organism does not merely consult the world; it has altered the world so that the world can later instruct it. The cognitive loop spans organism and environment. The self, in that moment, is not bounded by a membrane in the way we instinctively imagine. It is bounded by a set of causal dependencies that include the trace.

This view helps make sense of why decentralized systems can still appear purposeful. Purpose does not require a dictator; it requires a stable set of constraints that keep a system oriented toward certain outcomes. In brains, those constraints are implemented by circuitry shaped by evolution and learning. In slime molds, they are implemented by physiology and mechanics: oscillations, flow resistance, and growth rules that preserve useful pathways. In ant colonies, they are implemented by pheromone trails and local rules that, collectively, allocate labor and discover efficient routes. In fungal networks, they are implemented by branching growth and resource sharing across mycelium. None of these systems needs to *represent* the global plan in a single place in order to enact something that functions like a plan.

The harder question is whether calling this "mind" is a category mistake. One can be generous without being sloppy by separating two issues: intelligence and experience. Intelligence, in a broad and operational sense, is the capacity to solve problems: to reach food, avoid harm, and maintain organization under changing conditions. Experience—what it feels like to be something—is a different claim, and a much heavier one. Slime molds give powerful reasons to expand our definition of intelligence; they do not automatically settle questions about sentience. It is possible to accept that *Physarum* computes and learns in meaningful ways while remaining agnostic about whether there is anything it is like to be a slime mold.

Even without leaning on experience, the distributed-self idea has practical consequences. It influences what kinds of machines we build and how we interpret their failures. A centralized system can be brilliant and brittle; break the central hub and the system collapses. Distributed systems can be slower and messier, but they are often resilient. They can degrade gracefully. They can adapt locally without waiting for instruction. That is why engineers borrow ideas from biological networks when designing communication protocols, traffic routing, and robotic swarms. The goal is not to imitate slime molds aesthetically, but to exploit the same logic of local feedback yielding global order.

It also influences how we think about responsibility and control in human systems. A corporation, a market, a bureaucracy, a social media network: these are not organisms, but they can behave in organism-like ways. They pursue goals, maintain themselves, and resist reform. They also produce outcomes no one intended. When a system is distributed, asking "who decided?" can be the wrong question. The better question can be "what feedback loops made this outcome stable?" The slime mold's lesson is not a moral lesson, but it is a sobering analytic one: agency can be real without being owned.

There is an oddly intimate way to see this. Watch a slime mold faced with a choice between a risky shortcut and a safer longer route. Over time it may hedge: maintain a main artery on the safe route while sending thinner exploratory strands into the risky area. If conditions improve, the risky strand thickens; if conditions worsen, it fades. That hedging resembles a personality trait, a preference for caution or boldness. But in the slime mold it is not a trait stored in a mind; it is a balance of flows and reinforcements that can shift with hunger, moisture, and prior exposure. The self is not a fixed point. It is a dynamic compromise.

A compact way to hold these ideas together is to imagine intelligence as living on a spectrum defined by two axes. One axis is how centralized the decision-making is: does a system have a single controller, or does it distribute control across many parts? The other axis is problem-solving capacity: how flexibly and effectively does the system navigate challenges? Humans sit at high capacity and, for many tasks, relatively high centralization, though even human cognition relies heavily on distributed bodily and social scaffolding. Cephalopods complicate the picture: an octopus has a sophisticated brain but also extensive peripheral control in its arms, which can act semi-independently. Artificial intelligence can land in different regions depending on the architecture: a single model making decisions centrally, or a distributed swarm of simple agents. *Physarum* occupies an unusual quadrant: impressive problem-solving for its simplicity, achieved with strikingly low centralization.

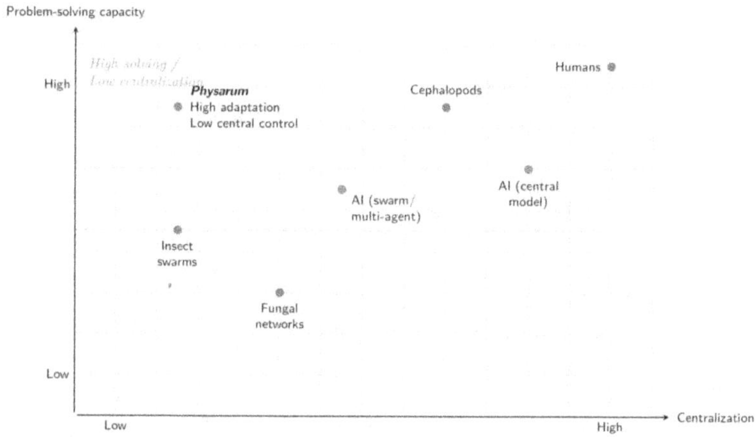

Seeing the slime mold placed among these other systems does not diminish human intelligence; it clarifies it. A mind like ours is one way to solve the problem of staying organized in a turbulent world. A slime mold is another way. The surprise is not that the slime mold is like us. The surprise is that coherent agency can arise when the agent is spread out, when the memory is partly in tubes and trails, and when the boundary of the self is drawn by ongoing coordination rather than by a single commanding seat.

If a self is what holds together in the face of disturbance, then a slime mold is a kind of self made of concrete particulars: dampness, pressure waves, chemical gradients, and the slow thickening of a tube that has earned its place. It is difficult to watch that stubborn persistence without feeling that "mind" is less a possession and more a way of being organized.

www.ingramcontent.com/pod-product-compliance
Lightning Source LLC
LaVergne TN
LVHW022206200926
848481LV00006B/1868
* 9 7 9 8 8 9 6 6 5 2 9 1 5 *